BETWEEN
APE AND HUMAN

BETWEEN APE AND HUMAN

An Anthropologist on the
Trail of a Hidden Hominoid

GREGORY FORTH

PEGASUS BOOKS
NEW YORK LONDON

To protect their privacy, the names of all Flores Islanders mentioned in the text have been replaced by pseudonyms. For the same reason, the names of Lio villages are mentioned only where necessary.

BETWEEN APE AND HUMAN

Pegasus Books, Ltd.
148 West 37th Street, 13th Floor
New York, NY 10018

Copyright © 2022 by Gregory Forth

First Pegasus Books paperback edition March 2023
First Pegasus Books cloth edition May 2022

Interior design by Maria Fernandez

Library of Congress Cataloging-in-Publication Data is available.

ISBN: 978-1-63936-381-0

10 9 8 7 6 5 4 3 2 1

Printed in the United States of America
Distributed by Simon & Schuster
www.pegasusbooks.com

To the memory of my father, Leslie William Forth (1925–2021),

who passed away after a long, productive, and generous life,

and just a few weeks after this book was accepted for publication.

CONTENTS

1

APE-MEN OF FLORES ISLAND

I n the far reaches of the Indonesian archipelago lies Flores (8.6574° S, 121.0794° E), a long narrow island of high mountains, precipitous cliffs, and deep ravines. The climate is tropical. Rain is seasonal and is heavy when rain-bearing "west monsoon" winds blow from October to April—definitely the wet season in this part of the archipelago. In southern regions, though, and in the high mountain forests that cover much of the island's interior, rain can fall in any month. By contrast, Flores's north coast, including a narrow coastal plain that runs along much of the island, enjoys far less rain, is less fertile, and reveals a semiarid landscape. Thus it is only certain parts of the island that live up to the name Flores, "Flowers," which early Portuguese navigators took from Tanjung Bunga, "Cape of Flowers," a name given by their Malay-speaking pilots to the island's eastern and, ironically, not particularly verdant extremity. (Why precisely they called the cape "flowery" remains a mystery.)

Like the rest of Indonesia, Flores belongs to the famous "ring of fire," the great chain of volcanoes that encircles the Pacific Ocean. Even by Indonesian standards, Flores is highly volcanic—another factor conducive to plant fertility. Yet contrary to what volcanoes, monsoon rains, and tropical forests might suggest, native animal life is neither rich nor varied. Because of the island's location well east of Wallace's Line—named after Alfred Wallace, the renowned naturalist and, with Darwin, cofounder of evolutionary theory—Flores falls within the Australasian zoogeographical region and is poor in mammal species. (Australia is obviously located in the same region, but Flores lacks the marsupials found on the great island continent.) Native mammals include mostly rats and bats. So most of the larger mammals now found on the island, including both wild animals (monkeys, exclusively *Macaca fascicularis*, or long-tailed macaques; porcupines; civets; and deer) and domestic species (pigs, water buffalo, goats, horses, dogs, and cats) were brought to Flores by human immigrants from the north or west—the earliest presumably coming in outrigger canoes.

While not short of birds and reptiles (including crocodiles and at least four species of venomous snakes), Flores therefore lacks—indeed has never had—such large animals as the tigers, leopards, bears, elephants, and rhinoceroses characteristic of larger western Indonesian islands and mainland Southeast Asia. For what comes later it's important to mention that Flores has never had apes either—those large, tailless primates represented by orangutans on Borneo and Sumatra and by gibbons on Borneo, Sumatra, Java, and Bali.

But what Flores lacks in animal variety it certainly makes up for in the remarkable character of several native species. Among these is the world's largest rat, the aptly named Flores giant rat (*Papagomys armandvillei*), a ferocious, mostly tree-dwelling beast that, including the tail, grows up to 80 centimeters long (or over 2.5 feet). The rat is an "island endemic"—meaning it is found only on Flores (see figure 1.1). And almost endemic is the world's largest lizard, the Komodo dragon (*Varanus komodoensis*), a voracious carnivore and ambush predator with a venomous bite that can attain a length of more than 3 meters (nearly 10 feet). Other than Flores, the dragon occurs only on its namesake island of Komodo and other small islands, all immediately west of Flores.

Looking back into prehistory, I could also mention the evolution on Flores of now-extinct pygmy stegodons, elephant-like creatures no larger than a cow.

But the subject of this book is an even more remarkable and even less expected animal—or perhaps two animals of very similar kinds.

One is an extremely small-bodied fossil human named *Homo floresiensis* ("Flores human"). The species is known only from remains found in 2003 at Liang Bua ("Bua Cave") in western Flores. Standing little more than a meter (3 feet, 3 inches) tall—the height of a two- to three-year-old Western child—the tiny species quickly became nicknamed "the hobbit," after the Tolkien characters. In view of this hobbit's skeletal features—described as "archaic" by paleontologists and, in several respects, comparable to those of Australopithecines (prehistoric "southern apes" that lived from two to four million years ago) or even chimpanzees—another surprise was the species' extraordinarily recent dates. Initially the last known date for floresiensis was estimated at just twelve thousand years ago, or eighteen thousand years ago for the "type specimen," or "holotype," that is, the most complete of several skeletons found at the same site (see figure 1.2). More recently the date was revised to fifty to sixty thousand years ago, but in geological terms this is still remarkably young. The discovery caused a sensation in the scientific world and captured the attention of the general public as well. And in spite of initial controversy, research conducted after the discovery has confirmed that *Homo floresiensis* is a new species.

The other humanlike creature alluded to above, which for convenience I call an "ape-man," has yet to be scientifically identified. But one of several ethnolinguistically distinct groups that populate Flores Island, a people called Lio, claims these creatures are alive (if not well) in remote sections of their mountainous territory. In their own language, the Lio (pronounced "Lee-oh") name these ape-men "lai ho'a." They describe them as small—in fact about the same size as floresiensis—as walking upright on two legs, and as hairy-bodied or, at any rate, hairier than themselves. Lio also characterize the ape-men as cultureless—lacking tools, weapons, clothing, and even fire. Floresiensis, too, might have been hairy, though we shall never know. And while the question has yet to be settled, there's no firm evidence for the fossil species having used either fire or stone tools.

Although often described as taller or larger-bodied, humanlike creatures like these ape-men have been reported elsewhere on Flores Island, in Indonesia, and in other parts of the world. But adding to the intrigue of the lai ho'a is the fact that—unlike similar beings reported from other parts of Flores, which local people consider extinct—these ape-men are claimed to have survived to the present. In addition, individual Lio offer credible accounts of specimens they have seen, including eyewitness encounters dating from the 1960s to as recently as 2017 or 2018. And all this on an island that, for thousands of years, was home to an "archaic" species of *Homo* (the genus that includes *Homo sapiens*), which according to the reconstructions of paleoanthropologists—anthropologists who study prehistoric humans—seem to have differed hardly at all from the Lio ape-man.

So far as we know with any certainty, modern humans (or what paleoanthropologists more exactly call "anatomically modern humans") first reached Flores around eleven thousand years ago, at the end of the Pleistocene (or "Ice Age"). By contrast *Homo floresiensis* had been living on the island since around a hundred thousand years ago, and the species' ancestors may have arrived much earlier. Because dates are available only from a single site, when floresiensis disappeared—or even if it disappeared—is not known.

As for the ape-men, there's reason to believe they could be present-day descendants of floresiensis, and if so it could mean that this species still shares Flores with modern humans. Alternatively, species X (as the ape-men might also be called) could descend from a similarly small-sized species that reached Flores over a million years ago and that may have been the ancestor of *Homo floresiensis* as well—but through a different line. Yet again, the humanlike creatures Lio speak of could be purely imaginary. Which solution is best supported by the evidence is what this book is all about.

Before going any further it's important to distinguish two similar-sounding terms, including one found in my subtitle. Meaning humanlike, "hominoid" describes any creature that looks like a human but is not a human—or at least not a physically modern human. "Hominin," by contrast, refers to a group (or scientific "tribe") that includes all species of the genus *Homo* (thus *Homo sapiens*, *Homo erectus*, and *Homo floresiensis*) as well as the Australopithecines. The Australopithecines include several species of erect-standing, bipedal apes known only from Africa, of which one apparently later gave rise to the genus *Homo*. At the risk of complicating things further, hominins form part of a larger group called "hominids" (note the /d/ in place of the second /n/!)—the zoological family Hominidae, which also includes the great apes (chimpanzees, gorillas, and sometimes orangutans).

Mostly to vary the prose, I use "hominoid" (not "hominid" or "hominin") as an alternative to "ape-man." That said, I recognize that if the ape-men were relatives of floresiensis, they would be hominins too.

As for the most modern of hominins, *Homo sapiens*, our chief interest lies in the Lio people, briefly introduced just above. Like other Flores Islanders, Lio possess a fascinating and, in many respects, little explored indigenous culture. But for present purposes, their main importance is as the sole witnesses to the continuing existence of ape-men, or what are possibly non-sapiens hominins, on their island. Living some 300 to 400 kilometers east of the western Flores site where *Homo floresiensis* was discovered, Lio occupy the largest part of east central Flores, one of the island's most mountainous regions. Though Lio territory extends to Flores's north and south coasts, especially in the south the highlands begin just a kilometer or less from the sea. This is rugged country by any standard and, even with modern roads (and if one drove straight through), the journey from Flores's western extremity to Lio's western boundary can take fifteen hours.

The Indonesian government does not keep records for ethnic populations. As Lio people are divided between two neighboring administrative districts, their current number is not recorded, but according to the broadest definition of "Lio," it is likely around a hundred thousand. At present, though, most Lio live near the south coast or in other, mostly southern, regions traversed by the Trans-Flores Highway, which runs from one end of the island to the other. For a long time Lio inhabiting coastal regions have traveled locally by sea and engaged in sea fishing while also growing crops. But even today only a minority of men are primarily or exclusively fishermen, and permanent coastal settlements of any size are a recent development. As this might suggest—traditionally and, to a large extent, still today—Lio make their living as cultivators and occasional hunters. They also raise domestic animals, which until recently were used exclusively as animal sacrifices in indigenous rituals and as items of exchange (chiefly as bride price given by a man's family for a wife).

Not long ago Lio mainly practiced "slash and burn" (or "swidden") cultivation in highland gardens carved out of mountain forests, where they planted corn (maize), dry-field rice, millet, and other cereals and a variety of vegetables and tubers. The gardens were maintained for several years before plots were abandoned so the forest could grow back. In the 1930s and 1940s, part of the population began cultivating rice in permanent irrigated fields located in lower-lying places closer to the coast. About the same time, some Lio also abandoned highland villages to build settlements nearer to paddy fields and modern roads (initially built by Dutch colonialists, who first arrived on the island early in the twentieth century and left after Indonesian independence in 1945). Even so, many cultivators still plant highland gardens and continue to reside at higher elevations, often at a considerable distance from roads and the sites of other modern institutions. During the twentieth century Lio forests shrank due to human population expansion. Yet the highest mountains remain covered in jungle, not least in several sections where forests are protected by government order. And it is mainly in these regions that the Lio say people, very occasionally, encounter ape-men.

Understanding Ape-men and What People Say About Them

The ape-man is a figure I've reconstructed from the statements of numerous Lio people. I have never seen an ape-man, and partly for this reason some readers may want to dismiss any resemblance between these creatures and apparently long-extinct hominins as mere coincidence. One purpose of this book is to question such dismissal.

I can immediately discount the possibility that the ape-man simply reflects local familiarity with the scientific discovery and reconstruction of *Homo floresiensis*. I was lucky. I first recorded physical descriptions of the Lio ape-man in July 2003, a month before a team of paleoanthropologists came across the remains of floresiensis in August of 2003 and well over a year before the discovery was announced to the public in October 2004. Even after that time, very few Flores Islanders learned much about the discovery, and those that did (mainly educated people with access to modern media) accepted the opinion of Teuku Jacob, an Indonesian paleoanthropologist, who dismissed floresiensis as a deformed modern human and a seven-thousand-year-old ancestor of certain short-statured villagers currently living close to the floresiensis discovery site at Liang Bua.[1]

But even though ape-men are an indigenous idea, this doesn't mean they exist as flesh-and-blood creatures and thus a real natural species—as opposed to an imaginary being existing only in people's minds. In the first case, there's the further question of what sort of species. If not some sort of hominin—including, perhaps, a largely "hidden" group of modern humans—Lio statements about ape-men might reflect nonhuman animals, either already known (monkeys, for example) or a species scientifically undiscovered. Incredible as it may seem, most evidence points to a hominin other than *Homo sapiens*.

Anyone can dismiss as "mythical" anything whose existence has yet to be proven. But actually demonstrating that the ape-men are imaginary is no easy task. One might attempt to show that the thing's existence contradicts the laws of physics or principles of biological evolution as these are currently known. To be sure, some Lio make what sound like fantastic claims about ape-men—for example, that they are able to "disappear" or even "fly." Yet many other Lio do not, adhering to a thoroughly naturalistic depiction. And if it is supposed that such naturalistically represented things do not exist, for anthropologists there is still the question of why people think they do—indeed, why some give seemingly credible accounts of ape-man sightings.

I certainly don't claim to be able to fully prove the hominoids exist. But there's a simple way to test how likely it is that ape-men are completely fantastic. This is to compare them with what Lio say about more definitely imaginary and fantastical entities, specifically supernatural or spiritual beings, as I do in chapter 3.

Even after showing how ape-men differ radically from spirits, at least one possible physical feature reduces their credibility as beings comparable to scientifically recognized hominins. For the fact is that some Lio say ape-men possess a short tail. This claim I examine at some length (if the pun be excused) in the next chapter. However the tail might be explained, though, it should be stressed that not everything people ascribe to a creature, even supernatural abilities, needs to be accurate for the creature to exist. Among zoological kinds, apparently impossible traits often combine with straightforwardly credible features to disguise the identity of a real animal. Later we'll meet several animals that Lio describe as supernaturally powerful, but which are unquestionably real species. For present purposes, though, another apparently fantastic creature may serve to make the general point.

Lio call this creature *beku*. The beku is largely nocturnal. It has a head and face like a dog and a long bushy tail like a cat, and it lives in trees. Alternatively, people describe the beku as looking like a large bat, though it lacks wings. Females as well as males possess testicles. In fact, the creature grows an additional testicle every year until it attains a full complement of twelve. Thus complete, it climbs a tree and wails throughout the night. Then at sunrise it drops dead.

Had I not been familiar with variants of the name "beku" from other Indonesian languages, I would likely have taken this for an imaginary, supernatural being. In fact the creature is the palm civet (*Paradoxurus hermaphroditus*), as Lio themselves recognize. The civet does have a head closely resembling the head of a small dog or a flying fox, a large fruit bat. Also, the animal is mostly nocturnal and arboreal, and it really does wail. The counterfactual element is the multiple testicles. Yet even these have a basis in zoological reality, for the organs in question are actually scent glands that resemble testicles, and their possession by both females and males is reflected in the species' Latin name, *hermaphroditus*.[2]

Of course, similarly questionable aspects of the ape-man cannot be resolved in quite the same way. Whereas I have seen civets in Lio and other parts of Flores, I don't have a physical specimen of an ape-man to compare with what Lio say the hominoids look like. Nevertheless, as a cultural anthropologist—someone trained in the detailed study and interpretation of cultural traditions and social systems—and as an ethnozoologist (or folk zoologist), who explores local, nonacademic knowledge of animals, I am familiar with the variety of ways humans speak and think about animals. As an anthropologist, I'm especially alert to how particular statements may reflect the sorts of experiences people may have with animals—or, indeed, not have—and furthermore how these statements may reflect personal and social interests or connect with shared representations of a community (or what is commonly called the "cultural context"). And to this task I also bring a background in eastern Indonesian languages.

I need to say more about how I came to learn about ape-men. All my studies on Flores have been broadly "ethnographic," meaning that I investigate how people organize their social and spiritual lives, what they believe, and what they consider valid knowledge. And like all ethnographers, I've pursued this research by living with local people over long periods of time and conversing with them in languages in which they are fluent. My first stint of fieldwork on Flores was in 1984; before that I spent a full two years on the neighboring island of Sumba, where I completed my doctoral research. Since 1984 I've returned to Flores nineteen times, typically spending two to three months in the field, thus amounting to a total of some four to five years.

The bulk of my Flores research has been among the Nagé people of west central Flores, and it was among the Nagé (pronounced "Nah-gay") that I first heard about mystery hominoids on the island. Later, in 2003, I began new fieldwork among the Lio, and it was in that year that, quite by chance, I learned about the Lio ape-men. Focusing mainly on these hominoids and other local creatures, I continued my Lio research in 2005 and again during five successive years from 2014 to 2018. During these years I lived mostly in the eastern Lio district of Mego.

Being less fluent in Lio than I am in Nagé, a large part of my ethnographic conversations with Lio people—so informal and open-ended have they usually been that I hesitate to call them "interviews"—were conducted in the Indonesian national language (Bahasa Indonesia), in which nowadays virtually all Lio are fluent. I have been speaking Indonesian now for over forty-five years, since I started my fieldwork on Sumba in 1975. At the same time, the Lio language is closely related to Nagé, and I am familiar with much vocabulary, especially terminologies used in talking about physical and behavioral aspects of humans and animals, a knowledge that greatly benefited discussions about ape-men and much else.

As already indicated, my major objective in this book is to compare two sorts of "reconstructions." One sort is the reconstructions paleoanthropologists have produced of extinct or presumably extinct hominins. Often, these have been made on the basis of sparse physical evidence. For example, the initial evidence for the Denisovan hominins who lived some forty thousand years ago in Siberia included no more than the bones of a single little finger. Similarly, the remains of seven-hundred-thousand-year-old possible ancestors of *Homo floresiensis* consist only of fragments of a single mandible and several teeth.

The other sort of reconstruction, and the subject of this book, is of the Lio ape-man, and the evidence in this case is not mute physical objects but what people say. For the most part, then, what follows is a book of stories, including stories by people who claim to have seen either living ape-men or, in at least two instances, their corpses. Altogether, these take up four chapters (5–8), preceded by a chapter (4) that discusses the relatively few appearances ape-men make in Lio myths and legends.

Attributed to several different sources is the proposition that "there are no truths, only stories." A variant substitutes "interpretations" for "stories." My approach follows the spirit of both statements. No doubt some would want to exempt "scientific truths" from the principle and, in the context of the present inquiry, perhaps assert that all the talk in the world is not worth a single finger bone. But even tiny bones require an interpretation, and such interpretations are typically part of larger paleoanthropological stories—often as appealing for the cast of fully modern characters they portray, and the twists and turns of academic fates, as they are for what they tell us of humans or humanlike creatures in the past. In addition, scientific claims to truth are always provisional, necessarily being subject to falsification—unlike religious truths (or "God's truth").

Many people, it seems, think that when scientific propositions are shown to be wrong, this reflects badly on science. But on the contrary, when decisively proven wrong, or even when seriously cast in doubt, science is actually advanced rather than weakened. The unlikely discovery of *Homo floresiensis* definitely changed our understanding of hominin evolution. Physical evidence for a contemporary non-sapiens hominin—which, however unlikely it might seem, is not impossible in principle—would naturally shake things up further still.

For the time being, though, we have to make do with what fellow human beings tell us. In this respect the present investigation might be called an exercise in cryptozoology, the study of "hidden" or "undiscovered" animals. Of course, the ape-men are not undiscovered by the Lio. So in this context "undiscovered" can only mean not documented or recognized by scientists, something typically involving the apprehension, by people with the proper scientific qualifications and professional connections, of either a living specimen or other tangible physical evidence. (Tracks might do, but photographs and recordings of vocalizations are more controversial.) Usually, cryptozoologists don't possess such evidence but instead construct their arguments for the existence of scientifically unrecognized animals on statements by ordinary folk who claim to have seen one. And since I take what Lio say about apemen seriously—as possibly reflecting a real animal—then this study can be called cryptozoological.

At the same time, as an anthropologist I'm fully aware that verbal evidence of any kind can be fabricated, exaggerated, or simply mistaken. I'm mainly talking, of course, about statements by people who claimed to have seen ape-men. One can go some way toward assessing the probable veracity of putative sightings by considering not only the physical setting and people's descriptions of what they saw (does it sound like another animal—a monkey, say—or a tree stump or rock, or indeed, something dreamt?), but also the personality, character, and social status of witnesses, including what interest they may have in representing something in a particular way. Certainly I place less reliance on accounts by people of a supernatural bent, most notably dealers in magically powerful ape-man body parts or substances (a topic explored in chapter 3) and "people of power" (*ata bhisa*) or "people of skill (or knowledge)" (*ata mbe'o*). These are typically men who have a personal—one might also say a professional—interest in claiming experience with things unfamiliar to others. As they serve as spiritual healers, magicians, and sorcerers, I also call these men of power "mystical practitioners."

Obviously the most valuable sighting reports are those where two or more people claimed to have seen the same thing simultaneously, and I was fortunately able to record several such accounts. But even in these instances all or both witnesses could have been mistaken, or misremembered what they saw. So it may still be asked: how can I be sure that *anything* I heard reflects a natural, zoological reality, or a creature corresponding to a scientifically unknown species? The short answer is that I can't. Even so, several considerations reduce my uncertainty—and may do the same for readers.

I begin with dishonesty and deliberate deception. Deception by informants is an easy charge to level at anthropologists. Researchers can find individuals given to deception in any field setting. As already indicated, there are ways of identifying people, with reference to both personal traits and social position or interests, who are particularly inclined to dissimulate or exaggerate. But there are also reasons to believe that a good deal of what people tell any anthropologist is accurate—at least in people's honest understanding. Here I should stress that most of what people said about ape-men agreed with what others said. And where accounts diverged from common views of the hominoids, it was usually possible to find a reason why.

One reason for thinking that what most people say most of the time is not made up is that deliberate deception and fabrication require effort. It's far easier for people to either tell the truth (as they see it) or simply say they don't know—surely, the best way to get rid of a troublesome anthropologist. Not only that, fabrication requires a motive. Sometimes, people may tell mistruths to convey what they think an inquirer wants to hear. But it's anything but clear why Lio might have thought I was soliciting thoroughly naturalistic depictions of ape-men. In fact, it's more likely that some people were inclined to stress, for my benefit, the supernatural aspects of the hominoids. This is partly because people knew I was also interested in such things as local spirits, witches, and magical beliefs and practices. Another reason is that some people were interested in selling me ape-man relics or were individuals who claimed supernatural abilities (those "men of power" again). Significantly, though, the accounts I found most credible, including reports from putative eyewitnesses, came from men and women who fell into neither category.

Sometimes mistruths might be told as a way of "pulling someone's leg," particularly if the topic is sensitive and details are difficult to verify (the number of a person's sexual partners, for example). However, whereas a Westerner might tell someone they'd seen something like a hairy hominoid to fool them—by convincing them of something the speaker considers untrue—this wouldn't work among the Lio. For Lio regard their ape-men as real although rare animals, whose existence no one disputes. A North American parallel might be telling a person you'd seen, say, a wolverine (*Gulo gulo*), an animal even experienced outdoorsmen rarely encounter in the wild, because even if the claim were false, it would hardly constitute leg-pulling.

To take this any further requires a more general look into Lio culture and worldview. This is a topic I can't possibly treat comprehensively here. But I provide an overview of indigenous spiritual beliefs and practices in chapter 3, while elsewhere I touch on other aspects of their past and current lives. For the present it's important to note that, for a century now, Catholic missions have been active among Flores Islanders. Large-scale conversion to Christianity didn't begin until the 1960s, when the large majority of islanders converted following Indonesia's anti-Communist coup. But Lio still recognize spiritual beings belonging to their native religion and still perform sacrifices and other traditional rituals. Before the arrival of the missionaries, the Lio language was entirely oral, and Lio did not become familiar with writing until the church introduced elementary Western-style education. Catholic education continues, though many schools are now run by the state. Until recent decades, however, few cultivators advanced beyond six years of primary school, and it is only in recent years that Flores Islanders have become familiar with either television or other mass media.

Also important for what follows, linguistic evidence I discuss in a moment shows that Lio were familiar with ape-men long before the arrival of missionaries or other agents of colonialism. What's more, secondary-school curricula have little if anything to say about paleoanthropology or human evolution. So far as I could discover, textbooks—or, for that matter, any of the extremely few books of any sort that are locally available—do not include illustrations of prehistoric non-sapiens hominins that could influence local ideas about hominoids living on Flores. This is not surprising. In my experience educated Florenese Catholics, including Indonesian priests, are openly critical of Darwinian evolution, and ordinary Lio villagers are barely aware of it. Their indigenous cosmology contains nothing comparable to a view of present-day humans or animals having gradually evolved from physically different ancestors. And as already explained, there's no reason to suppose that their view of ape-men has in any way been shaped by the discovery of *Homo floresiensis*.

As fits their description as rarely-encountered creatures, Lio seldom speak of ape-men and they mention the hominoids less often than other remarkable beings, including witches and various sorts of spirits. One reason is that ape-men play no part in Lio religion or ritual life, notwithstanding the use of reputed hominoid body parts as magical relics. Some people don't know much about ape-men. Yet I never encountered anyone who thought the hominoids were not real. Having no stake in and little knowledge of modern biology or paleoanthropology, Lio would have no reason to doubt that a small bipedal hominoid combining features of humans and apes could exist in their territory.

Nor, by the same token, would anyone have reason to dispute their existence or, for that matter, to prove that they do exist.

For Lio people ape-men are thus part of the local environment as much as any other rare creature—for example, Komodo dragons. (Some might want to claim that spirits are part of the local environment in the same way, but as I later show, Lio consider spirits as a radically different kind of being from ape-men). In their conception of the hominoids as natural beings, however, Lio people differ from Westerners convinced of the existence of Bigfoot (Sasquatch) or lake monsters, and equally from others who tend to dismiss their possible existence. Unlike Bigfoot believers, no Lio ever goes looking for ape-men. This follows in part from their belief that anyone who deliberately searched for a hominoid would never find one, so that encounters with the creatures are always fortuitous. Interestingly, Lio apply the same principle to other rare animals, including some recognized by science as zoological species. But also unlike aficionados of better known "cryptids" (animals unrecognized by science), Lio would have no obvious motive—financial or otherwise—either for finding an ape-man specimen or for perpetrating a hoax. And because they are barely aware of modern scientific views—or any view that claims such creatures do not or cannot exist—they have no ideological interest in proving any "scientific establishment" wrong. Expressed another way, Lio cannot be said to fully "believe in" the existence of ape-men because they've never been exposed to any contrary view.

As this should suggest, Lio are also less biased than those scientists who reject the existence of mystery hominoids or, more specifically, the possibility of contemporary non-sapiens hominoids surviving to the present (see chapter 10). Other researchers (Jane Goodall, for example) are more open-minded. Even so, such attitudes remain implicit in the reaction of some academics and nonacademics to any knowledge of the natural world that has not received the seal of scientific approval. All I ask of readers of the present book is that they give the Lio people a fair hearing. This is not to suggest that Lio have a direct channel to reality. Yet skeptics should ask why Lio think ape-men exist, other than because of firsthand experience or the experiences of people they trust. As I later explain, cultural anthropologists might come up with other reasons, but for the moment I let the question stand.

As this is primarily a book of stories about people coming across creatures that sound like non-sapiens hominins, then many of the contents should prove fascinating in their own right. Throughout, I've remained conscious of the need to not let a good story get in the way of, if not "the truth," then more mundane interpretations of the hominoid subjects. Still, what I've found so arresting about ape-men—and what led me to write this book in the first place—is the naturalistic or realistic way people describe them. If accepted as probable or plausible accounts of real creatures, what Lio say about the hominoids presents a challenge to zoology and paleoanthropology. If not, then a greater challenge falls to other disciplines concerned with human beings: to explain why people subscribe to the existence of entities with little or no basis in observable reality.

Of course, meeting either challenge is not straightforward. Not only in biology but equally in my own discipline of anthropology (and especially cultural anthropology), there is currently no body of theory that accommodates mystery hominoids, at least none that allows for their possible existence. If anthropology accepts beings intermediate between modern humans and apes, it does so only to the extent that they are treated either as a fantastic image (to be explained on mostly sociological or psychological grounds) or as species that became extinct tens or hundreds of thousands of years ago. In consequence, ape-men that are not imaginary or long dead are a topic without a discipline. But the concept of "story" may provide some way around this. In varying degrees stories can be accurate or inaccurate, authentic or fantastic, and it is ultimately for readers to decide into which category fall the stories of hominoids I explore below.

The Name of the Ape-man

I devote the rest of this chapter to the ape-man's Lio name: "lai ho'a." I am the first author to write at any length about lai ho'a.[3] A being recalling aspects of the ape-man receives a brief mention in anthropologist Eriko Aoki's doctoral thesis on western Lio. But that one is called *saga boko* ("short one, shorty") and is apparently conceived in a less naturalistic way.[4] Otherwise, "lai ho'a" appears only in the voluminous Lio-German dictionary published in 1933 by the missionary-linguist and ethnographer Paul Arndt.[5] This is significant. Arndt began compiling his dictionary in the 1920s, so the name has been in use for well over a hundred years and, in all likelihood, very much longer. And, since the Dutch established a colonial administration in Lio territory not long before 1920, we can be sure that the figure lai ho'a refers to does not reflect knowledge either of human paleontology or of Indonesian apes, which some Flores Islanders are now familiar with from travel to Borneo

Arndt translates lai ho'a simply as "evil spirit" (German "böser Geist"). However, he employs the same gloss for other things that, at the time, he did not fully understand or did not personally believe in—including headhunters still rumored to search for human heads for magical use in construction projects. (While this rumor is probably unfounded, the putative villains are in no way supernatural or spiritual; indeed, they sound all too human!) More insight into what lai ho'a refers to can be gleaned from several sample statements Arndt employs to illustrate uses of other entries in his dictionary. From these we learn that the creatures "closely resemble humans [in physical form]," that they are "small, tiny" and "thin;" that they walk or move in a peculiar manner; and that they are invulnerable to metal weapons.

All these attributions reflect not only Arndt's German translations but also Indonesian translations provided by several educated Lio with whom I discussed the statements in detail. I'll have more to say about these later. But for the moment we should note that nothing in Arndt's sample statements, not even invulnerability to wounding, suggests an "evil spirit." More importantly, the entries are anything but representative and, of course, provide nothing approaching a complete picture of what Lio understand ape-men to be. In fact, judging from Arndt's usual methods and those of other missionary-linguists working on Flores, the statements likely derive from a single text, probably a myth or legend, recorded and translated by Malay-speaking native assistants.

Since Arndt never mentions lai ho'a again in his later writings, including articles concerning Lio spiritual beliefs, he likely had little idea what "lai ho'a" referred to. And as he appears to have had little interest in or knowledge of local fauna (as indicated by numerous incorrect glosses of local animal names), he was probably not inclined to explore the matter further.

The name "lai ho'a" itself reveals little about its referent. The term comprises two words. According to Arndt's dictionary, "lai" has a variety of meanings, including "to resemble" and "to take, pick up, fetch." When reduplicated as "lai lai," it can mean "simple, short, concise," while in some Lio dialects "lai" names a dwarf palm tree. This last sense is interesting in view of the ape-men's short stature, yet it remains uncertain whether it bears on "lai" as a component of the hominoid's name.

The meaning of "ho'a," the other component, is even less certain. As Lio regularly describe ape-men's faces as monkeylike, it is noteworthy that "ho'a" resembles words for "monkey" in several languages of central Flores. But in Lio, "monkey" is always "ro'a," and I never heard "lai ro'a." Arndt gives no separate meaning for "ho'a" (which he transcribes as *hoä*). A word list for Endenese, a closely related language spoken to the west of Lio, gives "hoa" as the word for monkey.[6] But I've not been able to confirm this, so it could be incorrect. Arndt lists a similar-sounding term, which he writes as *ho'a*, as meaning "bare, naked." A speaker of the Jopu dialect of Lio also thought the second component of the ape-man's name meant "naked." This is suggestive insofar as Lio consistently describe ape-men as unclothed. Yet in other Lio dialects, the word for "naked" becomes *ko'a* (or *koä* in Arndt's transcription) and I never heard "lai ko'a." So, again, a decisive analysis of the name eludes us.

Asking people what features of ape-men the name might describe produced little result. Apart from "lai ho'a," eastern and central Lio appear to have no other name for the ape-men. Very occasionally they use "short one" (*saga boko*), the western Lio expression mentioned above. But it wasn't always clear that this is completely synonymous with "lai ho'a." As I later show, Lio are generally fearful of ape-men and, for this reason, are often reluctant to mention their name. So it's possible that some people use "short one" as a euphemism.

(Whether "lai ho'a" itself originated as a euphemism for an earlier name that has been lost is another question.)

When speaking Indonesian as well, Lio avoid pronouncing the ape-man's name by replacing it with euphemistic expressions meaning "that thing," or "that creature, or animal." Partly for the same reason, many people nowadays use the Indonesian word "kurcaci," actually meaning "elf, sprite," as an equivalent of the Lio name. This might suggest that ape-men are no more than imaginary beings after all. As I show in chapter 3, however, the matter is more complex, and the conclusion is mistaken. Pointing to a naturalistic image are two other Indonesian names for ape-men: "manusia hutan" or "orang hutan." Both terms translate as "forest people" (manusia means "human" whereas orang has more the sense of "person, people"), and Lio consider them synonymous. Readers may recognize "orang hutan" as the Malay source of English "orangutan," referring to the ape *Pongo pygmaeus*—in Malay and Bahasa Indonesia (the Malay-based national Indonesian language) also called "mawas."[7] How far the Lio usage links local ape-men with these Bornean and Sumatran apes is a question I consider later. For the time being it's sufficient to note that, while neither of the Indonesian terms literally translates lai ho'a, both reflect the humanlike (though not completely human) quality people attribute to the ape-man's physical form.

I must now explain my choice of "ape-man" as an English gloss of "lai ho'a." Once in scientific use, "ape-man" is no longer employed in anthropology—partly because it connotes such misleading notions as "missing links" and "humans descending from apes." But in contemporary English the name is problematic for quite another reason, since nowadays "man" is treated as an exclusive reference to human males.

I should therefore assure readers that Lio ape-men include females as well as males—as will be seen when I refer to "female ape-men" (rather than "ape-women"). This assurance may be insufficient for some, but all I can say is that, after much deliberation, I've decided to keep "ape-man" for two reasons. First, "ape-man" remains a widely known English term for a creature combining features of a physically modern human and an ape. So the expression is decidedly preferable to "ape-human" or "ape-people" (the intention of which is far less evident) and also better than any novel term I might invent. Yet a more important reason for adopting "ape-man" is because, more than any alternative I can think of—and the current gender specificity of "man" aside—it closely matches Lio descriptions of the hominoids as humanlike beings that reveal certain features of monkeys or apes—as we'll see in the following chapter.

The "ape" in "ape-man" also deserves comment, especially in regard to resemblances between lai ho'a and the fossil hominin *Homo floresiensis*. While classified as members of the genus *Homo* ("humans") like ourselves, floresiensis too displays many features of apes—both chimpanzees and Australopithecines. In fact, because of these features a number of specialists have questioned whether floresiensis should be included in *Homo*, not least because of the species' chimpanzee-sized brain. Peter Brown, a key member of the discovery team, proposed that the new species be assigned instead to a new genus to be named *Sundanthropus*. What's in a name? Well, "Sunda" is the name of the island chain that includes Flores (located more specifically in the eastern Lesser Sundas), while "anthropus" is the Greek word for "human." It might seem, then, that this alternative name still designates floresiensis as human. But actually it doesn't. Before the discovery of the fossil hominin, "*Homo*" applied mostly to humans like us, Neanderthals, and the now-upgraded "ape-men" (formerly named *Pithecanthropus*) *Homo erectus*. By paleontological and zoological convention, "anthropus," though Greek for "human," has always been applied to hominins that, in terms of their physical form and geological age, are intermediate between *Homo* (including ourselves) and both the great apes and more humanlike apes, such as the several species of Australopithecines.

Even the most adamant supporters of floresiensis's membership in the genus *Homo*—like the Australian leader of the discovery team, Mike Morwood[8]— would not deny that the species is far more "ape-like" than other members of the genus. And inasmuch as Lio people describe their local hominoids as being halfway between humans and apes, and therefore as "ape-men," we can see all the more clearly how they resemble floresiensis. While I was in the field, one of my Lio hosts once came across a reconstruction of *Homo floresiensis* in a book I'd brought with me. (This was the well-known painting by Peter Schouten. I didn't deliberately show it to them or anyone else because I didn't want to put ideas in people's heads.) The person identified the picture straightaway as depicting a lai ho'a (ape-man). Yet nowadays some Lio also apply this indigenous name to exotic apes like gibbons and orangutans. So for this reason alone there's obviously a lot more to say about what local ape-men might be.

PART I

WHAT PEOPLE SAY ABOUT APE-MEN

2

APE-MEN AS NATURAL CREATURES

What do Lio generally understand by "lai ho'a" (ape-man), and how far does what they say about how the hominoids look and behave add up to a coherent image of a natural, and hence credible, creature? In this chapter I focus on what I call "popular accounts," descriptions by people who had heard about ape-men but said they'd never seen one. Only incidentally do I mention reports by eyewitnesses or second-hand accounts relating experiences of putative eyewitnesses I was unable to interview directly (usually because they were already dead). The popular accounts provide a standard against which we can later compare reports of individual sightings and determine how much eyewitness descriptions correspond to or diverge from popular ideas.

Lio descriptions of ape-men are hardly uniform. Variation partly reflects the fact that individuals possess different degrees of knowledge of the creatures. By contrast, with eyewitness accounts a more likely factor is actual differences in size and other bodily features, corresponding to differences in level of maturity and sex we should expect in a natural species. Certainly, if either eyewitnesses or people in general offered only boilerplate depictions of ape-men, this would count against their zoological reality. For such depictions are more typical of spirits and other obviously fantastic creatures. (Think, for example, of unicorns and dragons.)

Most Lio were able to give some description of ape-men, though some could say little more than that they looked like small humans or bipedal monkeys. As this should suggest, some people describe them as more humanlike, others as more monkeylike (or since Flores Islanders don't distinguish apes from monkeys, we could alternatively say "ape-like"). But as we'll see later, reports by eyewitnesses give the impression of a decidedly more humanlike creature than do many popular accounts. A few people declined to describe the creatures on the grounds that they'd never seen one. As fits with the local view of the hominoids as extremely rare, however, the majority who did respond also said they had never seen an ape-man. Some people mentioned dead parents and grandparents as the source of their knowledge. Whether any of these sources ever encountered ape-men themselves we cannot know. Nor can we know how far non-eyewitness descriptions were derived, directly or indirectly, from what they'd heard from contemporary observers. Neither traditionally nor at present do Lio depict ape-men in paintings, carvings, decorated textiles, or any other form of graphic art. But ape-men are by no means peculiar in this respect, for Lio portray hardly any of the animals they know in visual media.

Lio people speak of ape-men as a kind of animal, at the same time remarking on ways they resemble humans—including erect posture and bipedal locomotion. With a height of a meter (3 feet, 3 inches) or more, they are also considerably larger than the far more familiar local monkeys: the aptly named long-tailed macaques, large specimens of which attain a head and body length of around 50 centimeters (1 foot, 8 inches).[1]

Older Lio—people over sixty-five, say—seem to conceive of ape-men as more humanlike, whereas younger people tend to consider them more "monkeylike." Assuming what older people say reflects an older view, this agrees with entries in Arndt's 1933 dictionary that describe ape-men as looking like humans. As already mentioned, Lio familiar with western Indonesian apes, from visiting Borneo or Malaysia for temporary work in plantations, apply the name "lai ho'a" (ape-man) to gibbons and orangutans. So based on their experience of these apes, it's possible that some people—especially younger people, who make up the majority of migrant laborers—have come to imagine Florenese hominoids as more simian than past generations did. That said, both young and old assured me that their parents or grandparents described ape-men as hairy-bodied and tailless, so these features cannot derive solely from more recent knowledge of exotic apes.

Ape-man Physical Form

Details of the hominoids' appearance come from accounts provided by 112 individuals. Of these, 32 were eyewitness reports, while 73 were descriptions by people who denied any personal experience of the hominoids. The remaining 7 come from secondhand reports by people recounting experiences of others. (These three categories are not quite mutually exclusive. Before relating his encounter, one eyewitness offered a general description of ape-men, as did four narrators of secondhand reports. So the total of 73 non-eyewitnesses includes a few people in other categories as well.)

I begin with height and size. People able to estimate ape-man height—and not everyone was—expressed this either verbally, in metric measurements (with which most Lio nowadays are familiar), or by indicating with a hand held above the ground. In seven instances, people simply compared ape-man height to that of a particular child (young children are ubiquitous in Lio settlements), whom I was often allowed to measure. Interestingly, where a child's actual height was compared to a prior verbal estimate or manual indication, the actual height was usually greater, and many of the tallest heights belonged to children I was able to measure.

Of all non-eyewitness descriptions that specified height, over half (56 percent, to be exact) gave figures of around one meter (3 feet, 3 inches). The highest was 1.2 meters (3 feet, 11 inches). If to these are added estimates of between 70 and 80 centimeters (2 feet, 4 inches), then nearly three-quarters

of respondents thought ape-men stood over 70 centimeters. That means a
minority offered estimates of around 60 centimeters (2 feet) or less, with two
people giving figures of between 30 and 40 centimeters (1 foot and 1 foot,
4 inches). What would account for such low estimates is uncertain. But a likely
explanation is confusion with certain media figures (usually named with the
Indonesian term "kurcaci"), as I explain in the next chapter.[2]

To get an idea of what these various heights look like, the reader should
refer to figure 2.1. Just two people specified different heights for males and
females, with the males being only slightly taller. One man mentioned this
while relating a story about an ancestor (see Nuwa in chapter 5); the other
was an eyewitness describing an encounter with a male-female pair (see Lalu
in chapter 6). The tallest height I recorded—around 1.45 meters (4 feet,
9 inches)—was from a secondhand story. As we'll later see, secondhand reports
included, on average, higher estimates than did accounts by either eyewitnesses
or non-eyewitnesses. Nevertheless, it's clear that Lio consider ape-men to be
far smaller than themselves—and therefore closer in size to local monkeys.
Although generally small, with an average adult male height of not much more
than 1.6 meters (5 feet, 3 inches), few Lio men stand just 1.5 meters (4 feet,
11 inches). And, with the rare exception of achondroplastic and proportionate
dwarfs (formerly known as "midgets"), an adult human height of around one
meter is unheard of.[3]

Lio descriptions included few details of ape-man build. Two sample sentences in Arndt's 1933 dictionary refer to ape-men as "small" and "thin" or
"scrawny." (Here we should note that when talking about a creature's size,
Lio words for "small" and "large" refer more to width, girth, or bulk than to
length or height.) Two people described ape-men as "sturdily built" or "larger
[that is, broader] than humans" whereas another two said they were thin or
very thin. According to yet another two accounts, the hominoids are about
the size of a dog, meaning as bulky (or "meaty") as local dogs. Canines on
Flores are typically the size of a small hound or terrier; they are also slender
if not scrawny, although hunting dogs can be quite sturdy. Statements by
non-eyewitnesses about ape-man build, therefore, neither clearly confirm

nor contradict Arndt's description. But as we'll later see, both eyewitness and secondhand accounts may lend Arndt a bit more support.

Whatever might be made of people's views on ape-man size, Lio invariably describe the creatures as strong. As people often remarked, despite being smaller, they are far stronger than humans. Interestingly enough, this assessment applies to orangutans and chimpanzees—but I should remind readers that, according to the scientific consensus, neither orangutans nor any other kind of ape occurs on any Indonesian island east of Wallace's Line, that is, outside of Sumatra, Java, Bali, and Borneo.

Whereas Arndt's 1933 dictionary says nothing about ape-man body hair, Lio almost invariably describe the hominoids as hairier than themselves. At the same time Lio are, on the whole, less hirsute than Europeans. Of 51 popular accounts that mentioned body hair, the majority described the hominoids as either "hairy-bodied" (19 of 51) or "hairy like a monkey" (11). Another 11 described the body hair as denser, coarser, or longer than humans' but sparser, finer, or shorter than monkeys. Actually, "hairy-bodied" and "hairy like a monkey" are virtually synonymous, while "hairy-bodied" can also encompass the third sort of description. Note also that "hairy like a monkey"—an evaluation I've heard applied to relatively hirsute but otherwise normal humans both on Flores and in the West—could be an exaggeration. Few people (7 of 51) claimed the ape-men had no body hair, or had never heard anyone mention this, or said they were no hairier than humans. (The remaining responses were incomplete or inconsistent.) But "no body hair" might also mean "no hairier than humans"—rather than, say, as bare as the proverbial baby's bottom.

Several Lio who portrayed ape-men's hairiness as intermediate between humans and monkeys provided further details. Koba, an elderly man we'll meet again later, described the hair as "not too dense but thicker than a human's and less thick than a monkey's." Tangi, a younger man familiar with Bornean apes, put it this way: "[the hominoids'] hair is sparse like a human's, but longer, and not as dense as an orangutan's." A couple of people compared ape-man body hair to that of a dog, at the same time offering other descriptions (hairy-bodied or hairy like a monkey). Lio dogs, we should note, are uniformly short-haired, and the hair is typically sparse.

Usually Lio don't distinguish ape-man sexes with reference to body hair, or they describe the sexes as equally hairy. Just one older man, Nuwa, claimed only the males were hairy-bodied; the females, he said, were no hairier than humans. Yet another thought the sexes differed but was not sure how. Although less mature specimens might be expected to be less hairy, no one expressly mentioned this.

In contrast to eyewitness and secondhand reports, non-eyewitnesses barely mentioned parts of ape-men's bodies that were hairier than other parts. Just one man described the hominoids as hairier on the chest than elsewhere, at the same time vaguely indicating that certain (unspecified) other parts were hairless. Four people referred to facial hair. One described "much facial hair," another "mustaches (or beards)" like an old person, and the third, "hair on the cheeks and perhaps the sides of the head." According to the fourth, ape-man faces are "hairless, like a human's."

Just one non-eyewitness, an elder named Goda, mentioned pubic hair, which he described as thicker and more abundant than other body hair. On another occasion, Goda characterized ape-man body hair as growing or pointing upward, in contrast to both monkey and human hair, which points downward. This, however, suggests a fantastical inversion of a sort common in supernatural beliefs, and another man denied it, attributing upward-growing hair instead to "vine mothers," a kind of spirit we'll meet in chapter 3.

In determining to what degree Lio conceive of ape-men as humanlike, a special interest lies in the head hair. Like other Indonesian languages, Lio has different words for head hair (*fu*) and body hair (*bua*), so unlike English, there is rarely any ambiguity as to what "hairy" means. All popular accounts that specifically mentioned head hair described this as short or no longer than the body hair. Some compared ape-men to monkeys in this respect, though Florenese monkeys (long-tailed macaques), in fact, have a patch of denser hair or a longer quiff or crest on the top of the head, which is usually of a darker color than other hair. As we'll later see, only eyewitnesses

reported ape-man head hair as being longer than the body hair. Bald hominoids were mentioned in only two secondhand accounts.

Nowadays, nearly all Lio men cut their hair short, whereas traditionally
both adult men and women let their hair grow long, sometimes down to the
waist (see figure 2.2). The only exceptions are people with naturally short hair
or tight curly hair. So even ape-man head hair growing to the shoulders, say,
would be shorter than that of many humans with uncut hair.

Just one man mentioned gender differences in head hair. Both male and
female ape-men, he claimed, have short hair, but the hair of females is somewhat longer. Though hair texture was not often specified, neither eyewitnesses,
non-eyewitnesses, nor narrators of secondhand accounts ever described ape-
man head hair as anything other than straight, that is, never curly or frizzy.
So it is noteworthy that curly or frizzy head hair is common among Florenese
humans (again see figure 2.2).

By all accounts, ape-man head hair is always the same color as the body hair.
The largest number of descriptions by non-eyewitnesses (8 of 21, or just under
40 percent) mentioned a color like monkey pelage or employed an Indonesian
word ("abu-abu"), literally meaning "like ash" and indicating a gray, grayish-
brown, or brownish-gray color. As monkey hair is usually of this color, the two
specifications are, for all intents and purposes, synonymous. Others mentioned
"dark" or "black" hair (both terms glossing Lio *mité*) or "red, reddish" (Lio
méra or *toro*) and, in one case, Indonesian "coklat" (i.e., "chocolate"), a word
that does service for colors English speakers would classify as "brown." In Lio,
lighter browns are usually expressed with words for "red" as are deep shades
of yellow or orange. Thus three non-eyewitnesses described colors I recorded
as "red or yellowish," "rusty orange or russet," and "yellowish (with gray or
brown markings)." When I first heard these specifications, I suspected possible
influence from images of orangutans—in English sometimes dubbed the "red
ape." Nevertheless, two eyewitnesses—one reporting a sighting in the 1960s,
before most Lio were familiar with orangutans—also mentioned reddish hair.

A few non-eyewitnesses mentioned hair of more than one color, or a basic
color marked with streaks or patches that were usually lighter: gray or white.

One man described dark brown hair mixed with gray, which he compared to my own sixty-four-year-old head hair. Lio recognize both monkey and ape-man hair as varying among individuals. But how much the hominoids differ from humans, either in this respect or in regard to hair color generally, is unclear, though human hair on Flores is almost invariably black. The exception, of course, is white- or gray-haired elderly people. Some Lio spoke of the hair of older hominoids, too, as "white" or "light-colored." So it's worth noting that graying—actually a process where individual strands of hair turn white—occurs in great apes as well as in humans, though at different stages in the life cycle.

Of people who described ape-men as varying in hair color, only an elderly man named Woda spoke of the colors as distinguishing what he, uniquely, said were two different "kinds" of hominoids. Standing about a meter tall (3 feet, 3 inches) and extremely rare, one has brown body hair with a white (or light) stripe or band on either side of the chest. The other, smaller and somewhat more common, sort stands around 50 centimeters (1 foot, 8 inches) and has grayish hair.

Despite these variations, all popular descriptions of ape-man body and head hair—and, as we'll see, eyewitness accounts as well—specified colors ranging from black and various shades of brown to red and gray. None were, for example, green, purple, or sky-blue pink. This is important, because the color range applies to the hair (or fur) of all mammals, including human beings. Unlike hair color, Lio invariably describe ape-man skin color as "black" (or "dark," *mité*), the term they apply to their own skin color, though they also recognize lighter and darker variants in local human complexions. Just one popular account and two eyewitness reports mentioned skin darker than local people's. Florenese monkeys, too, can have dark skin, but monkey skin is often pink or reddish.

So far, details of ape-man height and size and body and head hair point to a creature intermediate between modern humans and nonhuman primates. The same applies to descriptions of the hominoids' heads and faces. By all indications, the head and body show the same proportion as they do in humans. Given by Rawi, a reputed eyewitness we'll be hearing from several times again, the only numerical estimate I recorded described a head 10 centimeters (4 inches) wide and somewhat longer. Though this man claimed to have seen ape-men, in this context he appeared to be speaking in general terms. Comparable sizes included in eyewitness and secondhand accounts of particular specimens are discussed in later chapters. Ten centimeters indicates a head about four times the size of local monkeys, long-tailed macaques. Interestingly, it is also about the size of the cranium of the type specimen of the fossil hominin *Homo floresiensis*.

Many people characterized the ape-man's face as monkeylike. Others said it was more humanlike, or basically human ("almost the same as a person's") but nonetheless distinct from a human's and, more particularly, ugly. A few others described the hominoids as looking like "forest people" (Indonesian "manusia hutan" or "orang hutan"), which can refer to orangutans—a creature Lio, like other Indonesians, classify as a kind of "monkey." Because "monkey-faced" is a standard reference to human ugliness in Florenese languages,[4] references to ape-man ugliness and faces like monkeys (or apes) amount to much the same thing. And as monkey faces, too, are humanlike (as Lio recognize), there's also little difference between "humanlike," "monkeylike," and a third category discernible in Lio descriptions—"intermediate between a human and a monkey." We might note as well that, since younger and female orangutans can appear more humanlike than Florenese monkeys, this comparison similarly suggests something halfway between a human and an ape. Just three people (one an eyewitness) compared ape-men's faces to those of dogs. But in at least one instance the resemblance referred specifically to a somewhat prognathous jaw. (As we'll see in a moment, people in various parts of the world have sometimes characterized other humans as having faces or heads like dogs.)

Three popular descriptions and one eyewitness report further depicted apemen as resembling very old humans. As mentioned just above, another account described the hominoids as having facial hair like an old person. Elderly people on Flores are seldom plump and are often skinny. So a gaunt face, especially one so thin that the brow ridges and cheekbones become pronounced, can further suggest a simian appearance—as can thinning head hair and coarse facial hair on elderly women. And by the same token, an elderly appearance matches descriptions of ape-men as possessing a thin build.

Illustrating these several points is an incident involving an elderly high-lander, a man in his seventies, who visited me while I was lodging in another Lio village. His sojourn caused something of a stir among my hosts who, after he left, remarked—with considerable hilarity but with serious intent—how our guest, whom they further described as monkey-faced and ugly, looked like an ape-man. One man surreptitiously took a photo of the visitor with his cell phone and later shared it around. (A more flattering photograph I took appears as figure 2.3.) None of my hosts claimed they'd ever seen an actual ape-man, and it is unlikely that any of them would have mistaken the man for an actual specimen—unless, perhaps, they encountered him naked in a forest. Even so, the episode shows how Lio can perceive an elderly human, or at least one of a certain appearance, as resembling an ape-man. And this further supports the converse idea, that ape-men resemble old people.

Few descriptions of ape-men included particulars of specific facial features, and when they did, these were usually vague. While one secondhand account described "sunken" or deep-set eyes (apparently referring to eyes surrounded by thick ridges of bone, or toruses), all other details of the eyes came from eyewitnesses. Just six popular accounts specified ape-man teeth, distinguishing these as either humanlike or monkeylike about equally; two others mentioned canines, but how far this reflected an overall assessment of the teeth as mon-keylike was unclear. Apes and monkeys, of course, have larger and longer canines than do humans, especially the males. Details of the ears and nose were similarly unrevealing.

Two men stated that people only ever see male hominoids, and an eyewit-ness who claimed to have encountered a group of ape-men said that when people see specimens they always appear to be male. (All the same, no one I spoke to doubted that, to procreate, there must be females as well.) The idea might suggest that the sexes are not especially distinctive—indeed, they seem to be considered equally ugly—so the specification of males is the default.

On the other hand, the subjects of several sightings were described as female. Besides, all but one of the six non-eyewitness accounts that mentioned sex-specific genitalia or female breasts described these as like humans' (rather than like monkeys' or other animals'). The single exception was Dala, an elderly man who, while claiming to have observed ape-men more than once, gave no details of any single sighting and seemed to draw instead on a popular image of the hominoids as monkeylike. For example, he spoke of female genitals that are "visible from the back" and like a monkey's, and male genitals as like a dog's.

Another man said the nipples of female hominoids resembled a dog's. But everyone else, eyewitnesses as well as non-eyewitnesses, characterized the breasts as humanlike but small. Just one man mentioned "long" female breasts, while another spoke of a lactating female his deceased father claimed to have encountered as possessing enlarged breasts. If the hominoids' breasts are small, they definitely differ from otherwise similar and sometimes reputedly extinct hominoids reported from other parts of Flores, whose breasts are described as so pendulous they could sling them over their shoulders. Over-the-shoulder breasts are also a feature people in other parts of the world ascribe to mystery hominoids (such as the Himalayan yeti, the "wildmen" of China and central Asia, and the "mili mongga" of Sumba) or to members of other ethnic groups (including Hottentots and Patagonians).[5] But among Lio I came across the image just three times, all in reference to the previously mentioned female spirits called "vine mothers."

Of twelve accounts that expressly identified ape-men's arms or hands as either humanlike or monkeylike, eight described these as humanlike. In the other four cases it wasn't always clear whether "monkeylike" referred only or mostly to a smaller size. More importantly, though, no one described the hominoids' arms as extremely long or as longer than the legs—like those of monkeys or apes. In a few instances described as long, ape-men's feet may be a different matter, but as evidence for the feet comes mainly from eyewitness accounts, I leave this topic for later chapters. One non-eyewitness—Tangi, a man mentioned previously—said ape-man feet had three toes pointing forward and two backward. Although no one else mentioned anything like this, it somewhat reminded me of the misshapen feet of some humans that can result from walking barefoot over rugged terrain or up and down inclines, where the big toe and second toe especially are turned outward or sideways.[6]

A man named Tipa claimed the hominoids' feet are "reversed" or "turned backward." In various parts of the world, people attribute "reversed feet" to spirits, and Flores Islanders characterize the feet of monitor lizards (both the water monitor, *Varanus Salvador*, and the far larger Komodo dragon) in the same way. But as I was able to clarify, in reference to both monitors and ape-men, the description refers not to feet that are literally back-to-front—obviously an anatomical impossibility—but to feet turned to the side or splayed outward, like humans with "duck feet" (also called "quarter-to-three" feet). There's no indication that any of these characterizations are derived from observing footprints (a source of the same idea regarding monitor lizards). Lio say ape-man footprints are hardly ever seen, and when they are they simply look like prints left by a small human or a child.

Although just one eyewitness reported the feature, seven non-eyewitnesses described ape-men as possessing long fingernails. Léwa, an elderly man who provided one of the most detailed non-eyewitness accounts, denied this, describing the nails as short, like human nails. As the nails of monkeys, apes, and humans grow until they break or are cut or bitten off, the hominoids' long nails may well be imaginary. Underscoring this possibility, one man qualified the attribution of long fingernails, explaining that "no one who has observed ape-men has ever been able to see these nails." Of all secondhand reports, just one narrator mentioned long fingers or nails. One man claimed that a human-like figure on a metal pendant that had rather long fingers was an ape-man, but this later turned out to be incorrect—or at any rate an interpretation no one else supported.

The same man further reported that ape-men frightened people by pointing their nails at them as a threat. The reputed habit reminded me of statements made separately by two other men, one reporting on a specimen reputedly seen by his neighbor. Both claimed that erect-standing ape-men hold their arms in a flexed position with the elbows close to the body, or as I recorded after one man provided a demonstration, "in a sort of grasping posture, with the hands turned downward (as though limp-wristed)." As parents of young children will know, holding the forearms forward in this way is characteristic of human toddlers learning to walk. It's also characteristic of normally quadrupedal apes walking short distances on two legs. Although reported by just two informants, as we'll see later, the practice might also suggest correspondence to certain archaic features of the skeleton of *Homo floresiensis*—as could other relatively detailed depictions of ape-men's limbs and feet.

Ape-men with Tails?

As an account of a possibly real animal halfway between humans and apes, people reading this summary of popular descriptions of ape-men might, at this point, proclaim "so far so good." But we now encounter an apparent hitch, for some Lio said the creatures have tails. Tails are one of the few physical features of ape-men mentioned in Arndt's 1933 dictionary, where the missionary-lexicographer includes a sample sentence describing the lai ho'a as "like a human, except that it has a tail."[7] Abnormal extensions of the coccyx (or "tail bone") aside, a hominin or even an ape with a tail is an evolutionary impossibility. So if the claims were literally correct, we should have to conclude that ape-men are either monkeys or some other familiar animal.

That said, it's important to note that people in many parts of the world have falsely attributed tails to other humans including, interestingly enough, Europeans by other Europeans. Taller Africans have ascribed tails to pygmies. In recent times outsiders thought the similarly short-statured, dark-skinned natives of the Andaman Islands (located east of the Bay of Bengal, between Burma and Sumatra) had tails and, furthermore, heads like dogs. (Among the earliest writers to attribute dog-like heads to Andamaners was Marco Polo.) And closer to Flores—on the island of Sumbawa, immediately to the west—lowland Bimanese Muslims describe the Wawo highlanders of southeastern Sumbawa as possessing tails.[8]

It goes without saying that all humans thus characterized are as tailless as people anywhere, so verbally pinning tails on humans that lack these appendages is hardly novel. But for ape-men there is more to say. For a start, Lio disagree about the hominoids' tails. Well over half of non-eyewitnesses who addressed the issue (25 of 41) denied they had tails, with some rejecting the idea emphatically. Another man said he'd heard that only male ape-men had tails. The remaining 18 said they did have tails, yet all but 2 of these described the appendages as short or very short—for example, just "one or two centimeters" (0.39 or 0.8 inches) or "two to three" (0.8 to 1.2 inches) or "four or five" centimeters (1.5 or 2 inches) long. (Another 36 non-eyewitnesses either didn't mention tails or, when questioned, said they'd never heard of the appendages or had heard but did not know anything about them.) The two people who mentioned longer tails estimated these as 20 and 30 centimeters (8 inches and 1 foot). Yet these, too, are absolutely and proportionately far shorter than the tails of Florenese monkeys, which can exceed 60 centimeters (2 feet) and can be 50 percent longer than the body.

What's more, some people who attributed tails to ape-men said the appendages were not always visible. Koba, the elderly man mentioned earlier, said the tail was the same length as the body hair. Describing the tail as "just a bit of bone," another elder was more explicit about the hidden character of the appendage; "when [ape-men] stand erect," he claimed, "the tail can no longer be seen" because it is covered by body hair. On a separate occasion, the same point was made by a younger man. Yet another described ape-men's tails as "so short they are sometimes not noticed." And an eyewitness who denied that ape-men had tails observed that "[even] if they had very short tails these wouldn't be noticeable because of the hair."

Reports by eyewitnesses weigh even more heavily against tailed ape-men. The majority of observers who touched on the matter (16 of 23) described their objects as tailless or said they'd seen no tail. This leaves 7 people who did report seeing tails. But as we'll later see, details of these sightings (including length of the tail, the creature's overall size and form, and the place and circumstances of the encounter) indicate that what people actually saw was a monkey or another familiar animal. The only exception comes from a description of ape-men given by Rawi, a not always credible eyewitness, who said the tails of female hominoids were longer than the males'.

Nevertheless, while most Lio who attributed tails to ape-men described these as short or even invisible, the fact remains that a number of people evidently believe these tails exist, so we need to account for this. The best explanation is found in the way Lio classify living things, especially the sharp distinction they draw between "humans" (*ata* or *ata jie*; Indonesian "manusia") and "animals" (*ule age*; Indonesian "binatang," "hewan").[9] According to most people I asked, one difference is that only humans possess "souls" (*mae*). I should mention that, for Lio, this question is an unusual one, as their indigenous religion has no firm view on whether animals have souls. Even so, underscoring the status of ape-men as the most human of animals, Goda, a culturally conservative elder we met a little earlier, opined that the hominoids were one of just two animal kinds that had souls. The other was the dugong, or sea-cow (*Dugong dugon*), a marine mammal whose surprising resemblance to ape-men in local thought we'll learn about in the next chapter. But despite the ways ape-men are seen to resemble humans, no one I spoke to considered the hominoids—or sea-cows, for that matter—simply human. As I've already mentioned, regardless of their intermediate character in relation to the human-animal divide, most people describe them explicitly as a kind of animal.

So what does this have to do with tails? In the first place, when enumerating nonhuman animals the Lio language obliges speakers to add a "numeral classifier." For this purpose, Lio use their word for "tail" (*éko*), so when referring to three monkeys, for example, they say *ro'a éko telu*, or literally, "three tails of monkeys." The only comparable English usage is when we speak of so many "head" of cattle. Lio, by contrast, use "head" (*kolo*) as the classifier for humans; thus they speak, for instance, of "three heads of people" (*ata kolo telu*).

As creatures that, despite certain humanlike features, are ultimately nonhuman, ape-men should always be counted in tails. One man suggested that, because the hominoids combine features of humans and animals, neither "head" nor "tail" was appropriate, so one should just say *la'i ho'a telu* ("three ape-men"). But I never met anyone else who agreed with this. Lio enumerate nonhumans in tails because animals, or the large majority of animals, including mammals, do in fact possess tails, whereas humans do not—the coccyx, or what Lio, too, call the "tailbone" (*toko éko*) notwithstanding. So it's easy to see how the ape-man's tail emerges not simply from their being categorized as a type of animal but also from a linguistic convention that bestows on the hominoids a "terminological tail." Noteworthy here is the response of Mbaku, a man who was familiar with gibbons from a sojourn in Borneo, whom I asked whether these apes had tails. At first he was unsure. But he soon decided they must have tails since they are "monkeys, apes" (Indonesian "kera"). The Indonesian name does not distinguish the two kinds of primates, though we should recall that, in zoological fact, neither gibbons nor any other apes have tails.

As further suggested by people who described the ape-man's tail as very short, or hidden or invisible, the appendage therefore appears more essential than physical. Similarly revealing the Lio view of ape-men as animals is their use of sex terms, words that distinguish males from females. From questioning and observation I found most Lio employ the animal sex terms for the hominoids (*metu* for females—e.g., *jara metu*, "female horse, mare"—and *mosa* for males). Still, six people, all of whom classified ape-men as animals, considered the human sex terms (*ata fai*, "woman, girl" and *ata kaki*, "man, boy") more appropriate, justifying this by pointing to their erect posture and humanlike faces. Another three thought that either the animal or human sex terms could be used. Claiming that people should properly employ the animal terms, one man qualified this by suggesting that "when people observe ape-men and realize their resemblance to humans, they may instead use the human sex terms."

What, we might wonder, could account for this partial difference between the numeral classifiers and sex terms? Most likely it has to do with the fact that the contrast between the sexes is a difference within a species, marked by different external genitalia that resemble humans'—whereas, with just one exceptional view, both male and female ape-men have tails or they do not. All the same, of the nine people recognizing the human sex terms as the proper or alternative usages, only one (a man who said either these or the animal terms could be used) claimed ape-men had tails—and then only very short ones.

Whereas many Lio regard ape-men as animals lacking tails, they also speak of a minority of humans possessing such an appendage. I'm not talking about people with abnormally long tailbones, something that affects a small number of people everywhere. Rather, I refer to a belief, also found in more westerly parts of Flores, that people grow a short tail if they reach an extraordinarily advanced age. Some Lio recognize this similarity between ape-men and the very elderly. Referring to the hominoids' tails, one man described these as very short, just 1 to 2 centimeters (or less than an inch) "like a very old person's." The comparison immediately recalls the assessment of ape-men as looking like very old humans. In a similar vein, an older man named Somba spontaneously asserted that ape-men are "like people of former times," whom he added also had short tails. Lio claim people formerly lived to a very old age, whereas people nowadays rarely do. Also, "people of former times" means ancestors, and extremely old people are on their way to becoming ancestors, so the two ideas amount to much the same thing.

In a *Current Anthropology* article I showed that attributing tails to elderly people reflects a perception of very old humans as growing to physically resemble animals.[10] But here I cite the belief mainly to illustrate how an idea that finds no support in visible reality can apply to beings—*Homo sapiens* of an advanced age—who are hardly imaginary. We should also not forget how Europeans have historically represented members of other races or ethnic groups as possessing tails—either as an inference from or a way of underlining their perceived lack of humanity in other respects. Just as relevant is the false belief of many modern Westerners that chimpanzees and gorillas—like the previously mentioned gibbons and orangutans in the Lio view—possess tails, which is evidently connected with a nonscientific Western classification of these animals as kinds of "monkeys."[11] Personal evidence of this mistake is a large plush toy gorilla given to one of my grandchildren, which sports a tail 7 centimeters (2.75 inches) long—thus longer than most Lio ape-men, it seems!

How Ape-men Behave

The way any animal moves is obviously closely connected with its physical form, so it's best to begin discussing ape-man behavior with locomotion. Lio describe the hominoids as walking bipedally. But this needs qualification. According to his own translations, illustrative statements in Arndt's 1933 dictionary provide no clear picture of how ape-men walk. However, interpretations of the Lio sentences provided by three Lio assistants depicted the hominoids as "running or jumping about and quickly changing direction like a soccer player;" "moving about on the toes in a jumping (or hopping) fashion," and also "slowly moving with a hopping gait." Although the translators may have been influenced by their own ideas about the creatures, their interpretations nonetheless provide some insight into how ape-men move bipedally. As we'll see in later chapters (5 and 6), both eyewitness and secondhand accounts of individual sightings have the hominoids moving in similar ways. And not just that, skeletal evidence reviewed in chapter 10 indicates possible correspondence with the way that *Homo floresiensis* would have walked.

Briefly describing a creature he'd seen that he took to be an ape-man, the man named Tipa said the hominoid "walked badly," something he attributed to its feet being "turned backward" (or it having "duck feet"). Four other men described hominoids as slightly stooped or hunched over when they stood or walked, which may recall the idea that ape-men resemble elderly humans. Such a posture was also suggested by one man's interpretation of another of Arndt's sample sentences, though this might alternatively mean deliberately hunching over to appear small or inconspicuous.

Attributions like "walking by jumping" or "walking badly" are hardly precise. Even so, it is worth remarking that, among hominins (modern humans and other species of the genera *Homo* and *Australopithecus*) bipedal movement can be achieved in more than one way. Apparently conveying a more energetic form of locomotion, two eyewitnesses who mentioned "leaping" or "jumping" described hominoids moving in this manner while running away and retreating to higher ground. Another possible observer, the previously mentioned Rawi, also spoke of the creatures springing from a distance of about two meters (6 feet, 7 inches) as a form of attack.

While they of course leap from branch to branch or tree to tree, monkeys do not move bipedally in any of these ways. Further qualifying ape-man bipedalism were references to ape-men sometimes moving on four limbs. Obviously this is more suggestive of monkeys, yet Lio are well aware that monkeys are quadrupeds that only occasionally stand on two legs. In addition, proponents of occasional quadrupedalism described characteristically bipedal ape-men as moving on four limbs only in certain situations. As one man explained, "when ape-men enter a field to steal maize (corn) they walk on four limbs, but while still in the forest [their usual abode] they walk on two." Along the same lines, a second man—the previously mentioned Somba—described the hominoids as usually walking on two legs but going on four "like an animal" when disturbed by humans. He added, however, that when chased by dogs or people, ape-men will "stand erect and flee on two legs." Similarly, a third informant, referring to a hominoid he claimed he'd seen running away, depicted the creature as beginning its escape on four limbs but, after a few paces, standing up and running away on two. And yet another man said that, on hearing barking dogs, the normally bipedal hominoids will flee up inclines on four limbs to seek refuge at higher locations, where they resume an erect posture.

Most descriptions of ape-men as partly or occasionally quadrupedal (12 of 22) came from people who'd never seen one. Also, of eight or nine sightings of hominoids moving on all fours, all except one apparently involved monkeys (or, in at least one instance, a civet)—especially because all observers described their subjects as tailed and, in some instances, long-tailed. Certainly, then, some people sometimes mistake familiar animals, and especially long-tailed macaques, for ape-men. Such experiences could also be one source of the idea that the hominoids sometimes move on four limbs. Yet in view of the common understanding of ape-men as essentially bipedal animals, the idea that they occasionally engage in quadrupedal locomotion could also be notional in the same way as their possession of short or hidden tails. In fact, there's a more direct connection between the two ideas for, as mentioned earlier, in one opinion ape-men's tails are not visible when they stand erect, which of course would also mean when they walk bipedally. Note also that just as the tails are very short, so the four-limbed walking is always described as partial and occasional.

By no means did all popular accounts that mentioned quadrupedal loco-motion or tails specify both. This should not surprise us. If ape-men are animals, in light of their various human features, they are decidedly ambiguous animals. Like talking about them with the animal numeral classifiers (the "terminological tails") and the animal sex terms, short tails and occasional movement on all fours are alternative ways of expressing the ape-man's animal nature. Also, the minority view that ape-men are stooped—in other words imperfectly erect—or that they walk badly on two legs might be understood in the same way.

As eyewitness reports describe ape-men in single locations, most information on how and where they live necessarily comes from popular descriptions. Lio characterize ape-men as creatures of high mountain forests. Sometimes specific highland locations are mentioned as places where ape-men are most numerous. These include a region named Tana Wawo (literally "Land Above"), mountains called Pu'u Buti and Ndura, and the higher reaches of several other peaks. Some sightings are reported from locations on or near modern roads, but these roads mostly run through highland forest. To be sure, I did hear stories of ape-men appearing on the coast. But Lio claim they come down from their mountain retreats to lower-lying locations only occasionally. Also, most reports of ape-men near the coast, including most of the seven sightings mentioned just above, turned out to have reflected encounters with monkeys.

Put another way, Lio depict ape-men as inhabiting regions distant from human settlements, often in locations difficult for people to access, and living, as one man put it, in "places where people do not live." A partial exception are highland gardens and adjoining huts—temporary dwellings where cultivators overnight during parts of the year and where they store tools and produce. Typically surrounded by forest, the garden huts are a common setting for older stories of encounters with ape-men, as we'll see in later chapters.

Matching their ideas about where ape-men live, Lio say that, as the human population has expanded and people have started to clear forest higher upslope—developments that have accelerated during the last several decades—the hominoids have retreated to more elevated locations, including forested areas that now enjoy government protection. By the same token, Lio suggest that ape-men were more numerous in the past, and even that the creatures are now extinct. According to one idea, the hominoids became scarce after the series of earthquakes that followed the Flores tsunami of December 1992, which caused water sources to run dry and land to subside. Contradicting the suggestion that ape-men may now be extinct, however, are more recent sightings, the most recent of which, reported by a woman named Jata, dates to 2018 (see chapter 6). Just the same, Lio seem never to have regarded ape-men as common.

Apart from loss of territory due to human expansion, there is also the idea that ape-men are rare because they are inherently unprolific—or in the words of one man, they "take a long time to reproduce." Certainly if the hominoids were anything like present-day hunter-gatherers, or for that matter non-sapiens hominins, their fertility rate would be a good bit lower than that of cultivators like the Lio. The rate would be even lower if they were more like apes.[12] But whatever its cause, rarity is a complex phenomenon, and just because a species' numbers are low, this doesn't mean it's on the verge of extinction. Many mammals have small populations that have remained stable over long periods of time.[13] Also, the fact that local people seldom see an animal doesn't mean it is particularly rare. Indeed, its ability to remain "hidden" from humans may be an important factor in its survival.

Recalling the Western characterization of earlier humans as "cavemen," Lio often describe ape-men as living in highland caves. Only in one instance did I hear of coastal caves reputedly occupied by the hominoids, located in the north central part of Lio territory. Two elders more knowledgeable than most about ape-men stated separately that the creatures do not permanently occupy particular caves, thus implying a nomadic lifestyle comparable to that of human food collectors who regularly move between different sites to exploit different resources. Another man suggested the hominoids might sleep in trees, and a secondhand report described someone seeing a small ape-man high in a bishop's wood tree (a species with medicinal properties), stripping off bark and eating it. But although there is a vague reference in an origin myth to the creatures once "living in trees," nothing indicates a consistently arboreal existence. More generally, Lio characterize ape-men as good climbers, of steep inclines as well as trees.

Two people described ape-men as occasionally mixing with or appearing near troops of monkeys—a behavior also suggested by an eyewitness report discussed in chapter 6. The claim is probably the source of another idea, which I heard just once: ape-man hair obtained from a snare in which a hominoid was caught can be placed inside a cage-like trap used to catch monkeys, because the smaller primates will be attracted to the hair. Several myths I later review depict a group of ape-men occupying a single cave. But two elderly men denied that ape-men live in groups, with one adding that they mostly occur in pairs, including male-female pairs. As we'll see, most eyewitness accounts concerned individual specimens. However, as no sighting took place at a location known to be regularly inhabited or frequented by ape-men, from these accounts we can't learn much about the hominoids' social lives.

Although Lio do not credit ape-men with the ability to make or employ tools, weapons, or fire, a story I later recount describes two or three ape-men approaching a garden hut to warm themselves by a fire lit by the human occupant, and another two narratives suggest the same behavior. If ape-men did use simple tools of stone or other materials, it's quite conceivable that local people would not know about these or would not recognize them as tools. Whenever I asked people what ape-men eat, they mentioned mostly wild plants, including wild fruits, nuts, tubers, young leaves, and tree bark. In addition, Lio accuse the hominoids of stealing garden produce, mostly corn (maize) but also cultivated tubers, gourds, cucumbers, and occasionally papayas and bananas. Lio say ape-men never steal or consume uncooked rice.

Though I never heard of anyone actually observing ape-men committing such thefts, a secondhand report describes a hominoid walking bipedally

through a cornfield, pushing the stalks aside with its arms. Ape-men
stealing from cornfields may recall an early scene in the first "Planet of the
Apes" movie, where the thieves are culturally-degraded humans. But better
comparisons, from zoology and ethnography, include crop theft by chimpanzees and orangutans and reports of human foragers (African pygmies,
for example) taking produce from the fields of neighboring cultivators.[14]

Lio are only too aware that other creatures steal from gardens, most
notably monkeys. Yet ape-men and monkeys, they say, have distinctive
ways of consuming fresh corn. Ape-men reputedly pull open the leaves and
eat the kernels before resealing the cob, whereas monkeys break or knock
down the stems and carry the cobs away to eat. Three people described
ape-men climbing Arenga palms to steal the toddy (sap or juice) that tap-
pers collect in containers suspended from flower stalks. Since the juice
begins to ferment immediately after it is collected, the claim could suggest
that the hominoids sometimes become intoxicated. But whereas mystery
hominoids from other parts of Flores, and from other parts of the world,
have been described as getting drunk on toddy or other liquor, no one ever
mentioned drunken ape-men.[15] Besides, if ape-men do drink palm juice,
they are not the only consumers, for Lio recognize that palm civets (also
known in English as "toddy cats"), fruit bats, and certain nectar-feeding
birds will do so as well.

People don't usually describe ape-men as killing animals or eating meat.
The main exceptions are occasional references to the hominoids sneaking
into villages at night and stealing chickens or piglets. Lio also accuse ape-
men of surreptitiously entering dwellings and taking cooked rice, fish, and
hen's eggs (from nesting baskets kept in or near houses or field-huts). People
in coastal regions accuse the hominoids of stealing gutted fish left to dry
on beaches. But like stealing from gardens, the creatures are rarely if ever
caught in the act. Several times householders told me they'd found a hen's
egg with a tiny hole in the shell and the egg sucked dry, something they
seem always to interpret as the work of ape-men. Yet all these thefts are
more plausibly attributed to other creatures: rodents or sharp-clawed civets

in the case of the sucked eggs; cats, snakes, or monitor lizards in the case of chickens; and—in the case of small pigs—dogs, pythons, or indeed humans.

Lio usually don't describe ape-men as killing and eating wild animals. Three elders, all in their seventies, said they never did so. But Rawi, an equally elderly man, described the hominoids as consuming frogs caught at night from a particular stream, and occasionally catching junglefowl (*Gallus varius*)—a wild bird closely related to the domestic chicken. Because junglefowl are characteristically low fliers that fall prey to other creatures, neither of these activities, it's worth noting, would require tools or weapons.

No one ever mentioned ape-men eating crustaceans, insects, larvae, or mollusks (including slugs and freshwater and terrestrial snails). However, modern Florenese humans consume species from all these groups—as well as frogs, many birds, water monitors, other lizards, and large snakes—and Flores monkeys, whose alternative English name is "crab-eating macaques," consume crustaceans and shellfish.[16] So there's no reason to think that local ape-men would not do so as well. Similar hominoids from the neighboring island of Sumba and from other parts of the world are described as eating grubs. In fact, when the entire range of nonvertebrate animals is taken into account, there's no reason to think that Flores could not provide sufficient foods, both plant and animal, to support a small-bodied omnivorous forest-dweller.

With reference to several kinds of behavior, people speak of ape-men as active at night. As we'll see in part 3, over half of sighting reports concerned creatures encountered after dark or sometimes late into the night. Yet by no means do Lio describe the hominoids as exclusively nocturnal, and all other eyewitnesses, including the most credible, described sightings occurring in daylight. As befits the hominoids' liking for young maize, people describe ape-men as most active during the wet season, lasting from October to May or June in highland regions, and especially from December to February, when the corn ripens. But the creatures might be observed at any time of the year—as eyewitness reports further reveal—and Lio characterize sightings as more a matter of an observer's luck than the season. A complementary notion has ape-men favoring damp weather, drizzle, and overcast conditions, when cloud descends and high mountain forests become enshrouded in mist, a preference attributed to mystery hominoids in other parts of Indonesia.[17] Somba, a man mentioned earlier, was sure that ape-men don't like to go out in the rain because, he said, their feet become muddy, making it difficult for them to run away. The claim obviously doesn't fit well with ideas about hominoids preferring damp weather and the rainy season. Yet it may somehow relate to physical peculiarities of the ape-men's feet and the way they walk, matters I'll return to later.

The hominoids' ability to avoid detection when engaged in stealing suggests a particularly intelligent creature. So it's no surprise that several people—all, as it happens, older men—described ape-men as "crafty," the "cleverest of animals," or, in one account, even "more clever than human beings." At the same time, Lio speak of monkeys, too, as clever animals, even possessing an intelligence comparable to people. Providing an especially detailed account, Léwa, a previously introduced elder in his seventies, described ape-men as "natural animals" (Indonesian "binatang alam") that, nonetheless, look like humans, walk erect, and are more "powerful" or "capable" (Indonesian "jago") than other animals. Illustrating ape-man intelligence, Mbaku, a younger man, referred to an incident in which an ape-man was suspected of stealing one of a litter of piglets. The culprit, he said, would have come back the next night for another piglet, but probably realized that, after the initial theft, people would be on guard. Just as importantly, though, I never heard anyone describe ape-men as dull-witted—in contrast to similar figures recognized in other parts of Flores and on Sumba.[18] Again, it's interesting to compare Lio ape-men with *Homo floresiensis*. As neuroanthropologist Dean Falk has shown, although showing a cranial capacity of just over 400 cubic centimeters—about the size of a chimpanzee's—this hominin possessed enlarged frontal and temporal lobes and other features "consistent with capabilities for higher cognitive processing." As one recent commentator puts it, floresiensis "probably had at least similar mental abilities to ours."[19]

Apart from mental ability, Lio credit ape-men with keen senses. So sharp is their eyesight they can see people at great distances before people see them.

Referring to both their sensory abilities and intelligence, one man added that ape-men can "know in advance what humans are going to do." Probably mentioned most often, though, is their acute sense of smell. As several people remarked, a person can never encounter an ape-man unless they are downwind of the creature. But while ape-men are good at sniffing out humans, the reverse seems not to hold. When questioned, few people associated the creatures with any distinctive odor. Several mentioned a smell like moss, "something rotten," monkeys, a wild animal, or "wild" or "forest" people ("orang hutan," probably a reference to orangutans). Yet these responses seemed improvised, and there appears to be no general idea about a distinctive ape-man odor.

Although Lio say ape-men lack tools—in fact, material culture of any sort—their reputation as highly intelligent creatures raises the question of their ability to communicate vocally—in other words, their possession of a language. There is a widespread notion that ape-men will parrot the speech of humans they encounter. As we'll later see, ape-men also speak in Lio myths and legends. But then animals speak in myths and legends the world over, even while storytellers regard the same creatures as mute in ordinary life. If the Lio ape-men reflect a population of hominins (species of the same genus as *Homo sapiens*), it is not ruled out that they would be able to speak or imitate the speech of modern humans. Nevertheless, I found no consensus on whether the hominoids have a language of their own, let alone whether they can speak the Lio language, and only one sighting account suggested anything like two ape-men communicating with one another.

More often, Lio identify various nocturnal sounds with unseen ape-men, as fits with the idea that the hominoids are often active after dark. These include the sounds of whistling, whooshing, a person weeping, a pig squealing, and vocalizations similar to (though distinct from) monkeys. But the sounds Lio most often attribute to ape-men are described as resembling the cries of owls or other night birds and imitated with a drawn-out and high-pitched "hoo," "woo," or "ooo," often reduplicated (as "hoo hoo" and so on).

In fact, whistling, weeping, and squealing are equally attributable to Flo-renese owls. The calls of nocturnal birds can also bear an uncanny resemblance to human vocalizations. Hence an obvious conclusion is that owl-like sounds are more likely made by owls! Nonetheless, in the association of these sounds with ape-men, we might glimpse the origin of another idea—that ape-men mimic human speech. When guarding fields at night, cultivators regularly cry out "hoo" or "woo" to scare away wild pigs and other crop-raiding animals. People utter the same cries when nocturnally signaling to unseen human (or presumably human) others, as a way of revealing their own presence and implicitly asking others to identify themselves. But Lio say ape-men, too, will echo these cries, a claim illustrated in several stories and even one eyewitness account (chapters 5 and 6). So it could be that, from this reputed hominoid behavior, Lio have developed the widespread idea that ape-men like to repeat human words and phrases.

If ape-men are as clever as people say they are, it comes as no surprise that Lio describe the creatures as wary or afraid of humans. Despite claims that thieving ape-men enter settlements at night, Lio more often describe the hominoids as fleeing whenever they become aware of human presence. Ape-men are equally afraid of dogs—animals that regularly accompany people, men especially, whenever they leave their villages to hunt or visit distant gardens. I also heard of incidents of dogs savaging and killing ape-men or chasing the hominoids into trees. Besides humans and wild pigs, dogs, including small populations of feral dogs, are the only common animals that could conceivably harm a small-bodied hominoid. Even so, three eyewitnesses described ape-men as initially standing their ground and apparently showing no fear when confronted by either dogs or humans. Komodo dragons—carnivorous lizards that grow to up to 2 meters (6 feet, 6 inches) long in this part of Flores—would also pose a danger to tiny ape-men. But although the giant lizards occur in Lio territory, or did so until recently, the dragons are coastal creatures. Also, their numbers are very small, and during the twentieth century they became restricted to a few dry regions on Flores's north coast.

Just as ape-men are reputedly afraid of humans, people are afraid of them—as eyewitness descriptions of their reactions to hominoid encounters often revealed. Sources of this fear could include the ape-man's exceptional strength and the more fantastic claim that they are invulnerable to human weapons. Yet Lio don't usually characterize the hominoids as aggressive. In one opinion they will never attack or kill people unless someone "disturbs (or bothers)" them. As we'll see, two eyewitness reports described an ape-man throwing stones. But in neither instance was the target a human.

As several eyewitness accounts suggest, the creatures' physical and particularly facial appearance is probably more relevant to Lio fear of ape-men. In major respects (upright posture, bipedal locomotion), the hominoids are humanlike, but in other ways (hairy, naked bodies; ugly, monkeylike faces) they are obviously different from modern humans (*Homo sapiens*). Focusing on this difference, and sometimes enhancing it by attributing tails and occasional quadrupedalism to the hominoids, Lio classify ape-men as "animals." Yet this act of folk categorization—locating ape-men on one side of what for Lio (and probably for all humans) is the most radical contrast among living beings—hardly resolves their anomalous and indeterminate character. Being psychologically representable as neither fully human nor fully animal, ape-man appearance, and especially the ape-like visage (sometimes described as "terrifying" or "horrible"), remains uncanny. Add to this the experience of a humanlike creature appearing naked—a condition that Flores Islanders interpret as a sign of madness in humans and hence potential danger—and we can easily see how encountering such a creature could cause emotional trauma.

This interpretation finds support not only in individual reports of people becoming afraid upon encountering an ape-man but also in claims that a person might faint, become deranged, fall ill, or even die simply from looking at the creature. According to one report, a young man who'd seen an ape-man on Flores's north coast became seriously distressed and after the incident was unable to sleep alone for some time. As I discuss in chapter 7, Bara, a young girl who came across an ape-man near a highland garden, subsequently developed a high fever and hair loss. At the same time, Lio legends tell of people who received mystical benefits from ape-men, particularly by obtaining parts of the creatures' bodies that served them as powerful relics. As I later show, the hominoids are not distinctive in this respect. For Lio also attribute either mystical benefits or psychic harm to contact with both spiritual beings and certain rare but fully natural animal species. And they characterize these as similarly frightening.

Local descriptions of ape-man behavior—living in high mountain forests, avoiding humans, subsisting largely on plants, and even stealing from culti- vated fields—paint a naturalistic picture of the hominoids. So too does their depiction as small-bodied, physically strong, intelligent primates, standing up to a meter (3 feet, 3 inches) in height with naked bodies covered in hair. Individual accounts of ape-men, of course, vary. Yet they appear sufficiently uniform to indicate a single kind of creature—one that is distinct from local humans and familiar animal species but ambiguously reveals features of both.

Popular depictions of ape-men may convey a reasonably clear picture of what the hominoids look like and how they live. However, eyewitness reports, and some secondhand accounts as well, provide a more detailed, though no less credible, view. The contrast is important because it shows that eyewitness descriptions do not simply reflect a common stereotype, and so lends support to observers having actually seen something. That said, some ideas about ape-men, such as "reversed feet" and long fingernails, might suggest that they are, after all, largely or entirely imaginary beings. I explore this possibility in the following chapter.

3

APE-MEN AS SUPERNATURALS

Many people in the West regard mystery animals, or "cryptids" ("hidden" animals), as imaginary—or, at best, imaginary until proven otherwise. "Imaginary," of course, refers to things that have no counterpart in the world of "empirical" reality, which includes only things we're able to perceive using one or more of our senses, especially our sense of vision. It's best not to equate things that are imaginary with things that are not "real." Spiritual beings (including God) are not empirical beings. Yet people the world over believe spirits are real or, not worrying about the matter too much, take their existence for granted. One reason they might do so is the common belief that spiritual beings have substantial effects (providing bountiful harvests, for example, or causing illness and death). And, in that respect, "believers" can accept the reality of spirits while agreeing they are "imaginary" insofar as that they cannot be seen.

Proving that something is not imaginary can be done by producing a "specimen," a term that, like "species" (from Latin *specere*, "to see"), ultimately refers to something that one can see. We have no specimen of a living ape-man. Nevertheless, the absence of material evidence doesn't mean that a thing is entirely fantastical. Differently expressed, it is not possible to prove that something is imaginary, or nonexistent—except, perhaps, to the extent that descriptions of the thing depart from natural law, so that it appears entirely "supernatural" (that is, beyond or "above" nature). In the case of the ape-men, however, what we can do is compare what Lio say about these creatures with what they say about a class of beings that we can more definitely say is "imaginary"—although not unreal for Lio. I refer to purely supernatural or spiritual beings. Like people everywhere, Lio describe spirits as psychologically identical to humans—thinking, having intentions, possessing feelings, and so on. But in other ways, they recognize humans, other animals, and ape-men, too, as beings that are very different from spirits.

Forest Spirits, "Vine Mothers," and Witches

Actually, showing how, in the Lio view, ape-men differ radically from spirits, and how they belong with natural rather than supernatural beings, is not difficult. Although nowadays almost all Lio are either Christians or Muslims, they continue to believe in several kinds of indigenous spirits, including a divine creator or "supreme being." For Lio all these have the common property of being "only spirit" (or "spirit alone," *mae mésa*) and "lacking bodies" (*tebo iwa*). I must stress that these ideas and the standard phrases Lio use to express them cannot be attributed to the influence of Christianity or Islam. For the phrases appear, in the same form, in the writings of Paul Arndt, based on research he conducted in the 1920s, at the earliest stage of modern missionary activity on Flores.[1] Besides, Lio explained them to me in exactly the same way Arndt recorded them nearly a hundred years earlier, thus indicating the phrases were pre-existing idioms and not constructions of the missionary-anthropologist.

As traditional spirits lack bodies, it's no surprise that Lio conceive of these beings as immortal, neither dying nor giving birth, and normally invisible as well. By contrast, ape-men, like humans and other animals, certainly possess bodies and so reproduce and inevitably die. With a qualification I discuss later, hominoids can be seen by anyone, including by two or more people at the same time. As for their mortal character, people sometimes distinguish ape-men from spirits as beings that are "living, alive" (*muri*). Whether Lio would therefore agree that spirits are, in all respects, not "alive" is a fascinating question, but one that would take too long to resolve here. Suffice it to say that, for them, ape-men—just like humans and other animals—are "alive" (*muri*) in the sense that they are "mortal"—a word deriving from Latin *mortalis* and referring to living beings that must ultimately die.

The natural character of ape-men may seem settled. Adding to their difference from spirits is the fact that the hominoids play no part in the rites of Lio indigenous religion and so are not in any way central to Lio life. Yet in certain respects the hominoids resemble three kinds of supernatural beings. So we need to give these a closer look.

The first are forest or earth spirits, called *nitu pa'i* (or simply *nitu*). Nitu spirits can appear in dreams, where they assume human form. Mystical practitioners (the "men of power" mentioned previously) sometimes claim to have seen nitu even while awake. But like other spirits, the forest spirits are otherwise deemed invisible, and Lio might detect their presence only from an uncanny feeling that may overcome a person while traveling alone in lonely places. As people further explained, travelers might suddenly become afraid for no apparent reason, begin to sweat or feel their body hair rising, or hear footsteps or the sound of wind when no wind is blowing. When finding oneself in a forest or another uninhabited place for the first time, one should, therefore, take ritual precautions. Otherwise the person may later fall ill—a sign of having been "greeted or accosted" (*mega*) by the forest spirits.

All the same, some people can gain benefits from these spirits. This happens when someone encounters a spirit in animal guise. Forest spirits can assume the form of various animals, including rarely seen freshwater turtles or giant land crabs (about which I'll have more to say below). When bestowing favors on people, however, Lio most often describe forest spirits as taking the form of a snake or eel—typically an exceptionally large or otherwise unusual specimen, or a specimen encountered in unusual circumstances or behaving oddly. Lio call such encounters "visions" or "revelations" (*saka sera*). Anyone experiencing such a revelation should thereafter make an offering near the spot where the creature was seen, otherwise they will likely fall ill. Alternatively, either before or after making the offering, the person may have a dream in which a forest spirit appears, conferring some special power or quality, either "good" or "bad."

Usually the spirit confers the power by way of a stone or, less often, a piece of wood or root. The place where the person finds the stone can be indicated in the dream but, however located, the beneficiary should keep the object as a protective charm. Although ultimately provided by a spirit, Lio often speak of these stones as coming from the body of the creature, most often a snake, by which the spirit manifests itself. Such charms can be used to heal others, or the recipient may enjoy success in agricultural or other economic pursuits. Though they may be obtained by any favored individual, Lio most often speak of mystical practitioners as owners of stones received from nitu spirits. If the power is "good," the recipient will be able to cure. But if "bad," the person gains the power to cause people illness or other affliction—in other words, to become a sorcerer or, what amounts to much the same thing, a witch. Lio also employ magical bezoar stones obtained from the carcasses of game animals—deer, wild pigs, porcupines, monkeys, and sea fish. (Bezoars are solid masses of indigestible material that accumulate in an animal's digestive tract.) But these stones, some of which I've seen, are not explicitly associated with spirits, and their powers are usually restricted to promoting success in hunting and fishing.

As fascinating as these ideas may be, I summarize them to make a particular point about ape-men. As we'll see, Lio can gain supernatural powers from the body parts of ape-men much as they can from forest spirits by way of encounters with snakes and other animals. Another similarity lies in the fact that Lio speak of nitu as spirits found in uninhabited places, especially high mountain forests, where they dwell in large trees and sometimes in caves. Mountain forests, of course, are also the abode of ape-men. But with the spirits, this spatial association further reflects the Lio view that nitu are the original and, in a sense, continuing owners of the land, displaced or replaced by present human occupants. When forest is cleared, therefore, cultivators must perform a rite offering food, drink, and other comestibles to the spirits. They also make offerings to spirits in annual agricultural ceremonies and, since Lio consider nitu to be the owners of game animals, in hunting rites as well. In fact, so closely are these spirits identified with the earth or land that, in these contexts, Lio call them "nitu of the earth" (*nitu pa'i tana watu*) and "people of the earth" (*ana tana watu*, a name that also refers to the first ancestors to settle in a region). Indeed, the spirits are also simply called *tana watu* (literally, "earth [and] stones"), a phrase used in formal speech for the earth itself—though we should be clear that Lio certainly recognize the difference between the physical earth and these earth spirits.

No such ideas apply to ape-men, and whenever I asked, Lio assured me that ape-men are in no way connected with bodiless nitu spirits. For example, they never describe ape-men as physical embodiments of the spirits. And rather than identifying them as anything like an aboriginal population (as they tend to describe the spirits), Lio speak of ape-men as having descended from ordinary humans, and so appearing later than humans, as we'll see in the next chapter. Also, whereas some people may occasionally encounter a forest spirit, in the form of either a human or a snake, Lio say this can only occur when a person is alone. By contrast, ape-men can be seen by two or more people at the same time.

If, despite certain parallels, ape-men have nothing to do with forest spirits, this may be less true of another kind of supernatural being. Called *ine léke*, these spirits are also encountered in uninhabited places. *Ine* means "mother" and is used when addressing women more generally. *Léke* is the name of a large forest vine. So I call these spirits "vine mothers." Whereas nitu include both males and females, true to the name "mother" Lio regard vine mothers as primarily, and perhaps essentially, female. Also, while nitu can manifest themselves visibly in various animal forms, the vine mothers, when they choose to reveal themselves—again, only to lone individuals—always assume the form of a human being.

As might be expected, vine mothers mostly take the form of women. Or at least they do so when they appear before men, for Lio say the spirits occasionally confront women and do so in the form of a man. In either case, vine mothers remain visible only for a short time—maybe less than a minute—after which they vanish. When a man encounters a vine mother, the spirit often appears as a beautiful but unfamiliar woman. So attractive are these vine mothers, a man will likely be seduced by them. But if they engage in sex, he is doomed and will shortly die. Alternatively, a vine mother might assume the form of the man's wife or other female relative (in one instance, a brother's wife). Looking just like his wife, the spirit will approach a man working alone in his field or in the forest, bringing his midday meal. And if he consumes this, his fate will be similarly sealed. Interestingly, Lio never mention having sex with the "false wife," and the beautiful women never bring food! A man can, however, ward off the advances of a vine mother by piercing her with a sharp metal object. The spirit will then be unable to move or return to her previous invisible condition, or as Lio also say, will be unable to "change or transform" (*mbale*).

One story I heard concerned a vine mother who approached a woman after she left her house in the early hours of the morning to relieve herself. The spirit assumed the guise of her elder brother and raped her. Later, the woman developed a fever and died the following night. It hardly needs stating that Lio ideas about vine mothers include a pronounced sexual element. By contrast, sexuality plays virtually no part in human relations with ape-men—though, as we'll see later, a few people suggested that some men in the past had married female hominoids. A myth I recorded in central Lio had an ancestor marrying a vine mother, but I never heard of any such matches in eastern Lio (the site of most of my fieldwork). Indeed, Lio in general describe these spirits as exclusively malevolent—and extremely dangerous. Forest spirits (nitu), too, can harm people, but they can also confer benefits on favored individuals. And as mentioned in the previous chapter, ape-men can do so as well.

A more obvious difference between ape-men and vine mothers lies in the fact that the commonest guise of these spirits is a beautiful woman, whereas people always describe ape-men as "monkey-faced" and "ugly." The Lio word for "ugly" (*re'e*), however, refers not just to physical but also to moral deformity and so additionally means "evil, malevolent, bad." As one Lio elder explained, vine mothers are ironic figures, for while they assume the outward guise of beautiful women (or at least humans that are not decidedly unattractive), the "real" faces of these spirits, which humans never see, are actually ugly. In a more general vein, Lio occasionally refer to vine mothers as "ugly (or evil) people" (*ata re'e*). These ideas seem to link vine mothers with the invariably ugly ape-men. But more to the point is the essential malignancy of the unseen character of these spirits. This invisible aspect is further reflected in another Lio belief, for people also say that the presence of a vine mother can sometimes be detected from a foul odor. Odors, of course, cannot be seen, so I would stress that, for Lio, vine mothers are essentially as "bodiless," or invisible, as are other spirits, and the physical forms in which they occasionally appear are never their authentic forms.

Before moving on, we should attend to another connection between vine mothers and the hominoids. Often, when discussing ape-men, people would remark "there are actually two (kinds)." This, it always turned out, referred to the hominoids and the vine mothers, thus suggesting that the two might be varieties of the same thing. What this more inclusive entity might be, people were unable to specify. Further inquiry revealed that Lio don't regard ape-men as a subclass of the vine mothers or vice-versa—as sharks (*iu*), for example, are classified as a kind of fish (*ika*).

Still, the remark suggests that some Lio understand ape-men as closely resembling vine mothers, or at least as resembling these more than other spirits. The resemblances are not difficult to identify. Both the hominoids and the spirits are beings that people see only rarely. What's more, when people do see them, they typically encounter both in lonely, uninhabited places and in specific humanlike forms. So it's most likely because of these similarities that people sometimes seem to confuse the two figures. Lio say the only way one can distinguish a vine mother in visible guise from a real human is by hands or feet with more or fewer than five fingers or toes (six or four, for example). A few people described ape-man digits in the same way, but other Lio denied this, explaining that they had likely got the hominoids mixed up with vine mothers.

In both their malevolent character and the fact that they outwardly look like ordinary human beings, vine mothers have more in common with another supernatural figure—the witch. One of the most noteworthy discoveries of cultural anthropology is the existence of a belief in witches in virtually all parts of the world. Far more than vine mothers, and like witches elsewhere, Lio witches (*ata polo*, "witch people," or simply *polo*) seriously challenge the human-spirit boundary. For people imagine them as humans that are empowered and inclined to act like "pure spirits" (*mae mésa*).

As seemingly ordinary people—and obviously natural beings—Lio witches are very often relatives or neighbors of the people who suspect or accuse them of being witches. Yet witches possess malign supernatural powers that inhere in an evil spirit (called *wera* or *ana wera*). This spirit is normally hidden inside the body of a witch. But, typically at night, it will leave the material body to cause sickness and death to the witch's enemies. Recalling forest spirits, a witch's *wera* can take the form of owls and other nocturnal birds. Speaking more generally, Lio describe witches—and not specifically their evil spirits—as temporarily transforming into larger animals. An elder named Demu related how his uncle, a man famous for identifying witches, once killed a male witch who had taken the form of a large dog by shooting it with an arrow. According to another report, a woman accused of being a witch was able to escape pursuers by turning into a monkey and climbing a tree. This story is particularly interesting because it depicts a person, in her ordinary human form, directly turning into an animal. Finally, I should mention that Lio witches are capable of flight—a common belief the world over. Among other things, Lio describe witches as abducting people and flying away with them.

Supernatural Beliefs About Ape-men

How ape-men differ from bodiless spirits and those "half spirits" called witches is already clear. But now I must introduce several supernatural bulls that might appear to wreak havoc in the ape-man's naturalistic china shop.

For the fact is that some Lio describe the hominoids as possessing fantastic powers otherwise identified with spirits. Among these are the ability to fly, to transform, and to disappear—all powers that Lio label with Indonesian words meaning "mystical" or "magical" ("gaib," "ajaib").

Actually, the damage to the metaphorical crockery is not nearly as bad as it seems. To begin with, none of these fantastic abilities is borne out by eyewitness accounts of ape-men. Nor do they receive any mention in Arndt's 1933 dictionary. But claims that hominoids can fly and the like still need to be unpacked and understood in the context of Lio language and thought and, more particularly, their ideas about more familiar animals and human beings.

Let's begin with flying, an especially curious form of locomotion for a wingless, bipedal primate. Apart from birds, bats, and insects, otherwise naturally described creatures credited with powers of flight include not just ape-men. For in addition to ultimately-human witches, Lio credit certain exceptional humans—those "men of power" again—with the ability to fly (as we'll see in the story of Leja, in chapter 4). On the other hand, by no means do all Lio believe ape-men can fly. The majority of non-eyewitness descriptions of hominoids (60 of 88) made no mention of flying, while of those that did, nearly an equal number said they could or could not fly. Another half dozen gave qualified responses—stating, for example, that ape-men could only "glide" a short distance from the ground.

As this last qualification may suggest, "flying" in Lio can mean several things. Both the Lio and Indonesian words for "fly" (*léla* and "terbang") can also refer to "moving rapidly" and "fleeing, running away"—as of course can "fly" and "flight" in English. As for rapid movement, even whales (named *léla ngai*) are said to "fly" (léla), meaning they can quickly traverse great distances at sea. (*Ngai*, "breath, to breathe," alludes to a local perception of their endurance as long-distance swimmers.) Indonesian "terbang" has the further sense of "to ascend rapidly," something Lio describe ape-men doing when they climb quickly up inclines. Interestingly, the Nagé, western neighbors of the Lio, use "fly" in this way when they portray a variety of wild pig, appropriately named the "eagle pig," as capable of leaping or propelling itself into tree branches and forest vines to escape pursuers.

I'll have more to say about "flying" later on. As for "gliding" (or "floating;" Indonesian "layang"), a few people said that when ape-men walk their feet appear not to touch the ground—by 10 to 20 centimeters (4 to 8 inches) in one estimate. One man said the hominoids seem to glide when they leap or spring, while another described ape-men as walking and running on the ground but appearing to "glide" like speedboats, which he'd seen during a visit to Borneo.

What these references to "gliding" mean is hardly clear. Some Lio claimed "gliding" actually applies to vine mothers rather than ape-men and, anyway, it's not certain that the statements all allude to the same thing. However, if the term describes a particular manner of bipedal locomotion, it could refer to what specialists in biomechanics call a "compliant gait," which contrasts to the "stiff" gait involved in modern human walking. Not just that, descriptions of ape-men walking with a hopping or jumping gait (mentioned in the previous chapter and further revealed in stories we'll later review, concerning encounters with individual ape-men) could ultimately refer to the same thing. "Compliant gait" doesn't translate easily into everyday English. But it mainly involves bending the knees and raising both legs while walking. For older readers, it might be compared to the "Groucho walk" executed by the best known of the Marx brothers. For younger readers, walking on skis or cross-country (as opposed to downhill) skiing may provide a more familiar comparison.[2]

More often than somehow moving with their feet off the ground, ape-men are described as "disappearing" or "vanishing" (Lio *bopa*, *bopa leda*; Indonesian "hilang"). This claim, too, is less fantastic than it may seem. The Lio terms can indeed refer to "vanishing" before one's eyes, or into thin air. But vanishing miraculously, or becoming invisible, is a power some people attribute to other human groups, like pygmies in Africa (as we'll see in chapter 9). In addition, like English "disappear," or even "vanish," the Lio terms are more often used in a mundane sense—to refer, for example, to an animal that quickly enters thick vegetation so it is no longer visible and people don't know where it went. Or to a person who's not been seen in a long time and can't be found.

More interesting than flying and disappearing (and also more commonly mentioned) is the idea that ape-men can change shape. About half of non-eyewitnesses questioned (19 of 37) thought the hominoids could transform; 6 people denied it, and the rest were unsure. Animals whose form ape-men might take include pigs, monkeys, civets, domestic and feral cats, and dogs. Answering questions about what "lai ho'a" referred to, four men said the name described a creature resembling a rat or mouse. Other people judged this incorrect, even absurd. But it's not difficult to see how the idea could have its source in the belief in shape-shifting. A few times people vaguely suggested that ape-men could also assume human form, but since they are regularly described as humanlike, this could refer to hominoids that had assumed another form subsequently resuming their original shape. Anyway, it's important to note that Lio consider an erect, bipedal hominoid to be the ape-men's original or authentic form. That is, they don't speak of this form as a shape sometimes assumed by an essentially different sort of being. Thus whenever people described the height of ape-men they implicitly referred to something standing erect, or occasionally in a humanlike sitting or squatting position.

Also implying transformation, a story related by a forty-year-old woman named Mbapu concerned a "lai ho'a" (ape-man) that used to enter her parents' house at night, in search of food. Once, after hearing a disturbance at the back of the house, she went to investigate and found that the intruder "had turned into a large black cat." One might simply conclude that the intruder was indeed a cat, but we need to take a closer look at the way Lio express such ideas in their own language. More revealing than the hominoid in cat form is another story, which concerned not an ape-man but another animal. One morning around 4:00 A.M., while it was still dark, Repu, a man in his fifties, was traveling by motorbike when he saw what he thought was a cow. When he got closer, however, the cow, he said, "turned into a rock"—evidently meaning that the animal "turned out to be" a rock.

Lio express the idea of transformation with the word *mbale*. As mentioned in reference to vine mother spirits being rendered incapable of vanishing (and thus unable to resume their original, invisible form), *mbale* further means "to appear and then disappear" or, with reference to normally invisible spirits, "to (temporarily) assume a visible guise." Yet as this example suggests, *mbale*, "to turn into," also covers senses that English speakers distinguish. Expressed another way, what we would consider a case of mistaken identity—realizing that something is not what we first thought—Lio understand as an instance of *mbale*, a broader concept that also encompasses what we would specify as "transformation." Without getting too involved in linguistic comparison, it's also worth mentioning that Lio *mbale* is related to numerous words in various Indonesian languages (including national Indonesian "balik") whose basic meaning is "to turn, turn over" (as in an overturned boat or bowl)—or as we might say in regard to the Lio concept of transformation, "to show (one's) other, or another, side."

Unlike Repu's story, only Mbapu's identification of the black cat as an ape-man concerns the notion that hominoids can turn into other animals, even though she acknowledged that she'd never actually seen her intruder in hominoid form. The same applies to a more explicit and more elaborate account I recorded in central Lio. Together with several companions, a young man named Siro was once fishing for crayfish at night in a mountain stream. Suddenly, they saw something coming toward them along a forest vine, which in Siro's words "took the form of a civet." This they found odd, partly because the animal was all white—evidently an albino or white morph of the palm civet (an animal introduced in chapter 1). White civets do occur on Flores but are uncommon. From the vine, the animal diverted to a nearby banyan tree. Yet when they pelted the tree to drive the civet away, what emerged—or rather, flew out—was an "eagle." As Siro explained, the "civet" had "changed shape." Like Mbapu and the cat, neither the man nor any of his companions saw anything like a hominoid. Nevertheless, he was quite sure that what they'd witnessed was a lai ho'a (ape-man).

I should add that these accounts are hardly typical of Lio ape-man sightings. For this reason I don't count them among the eyewitness reports reviewed in chapters 6 to 8, which include only accounts of subjects whose features observers described as hominoid, or at least primate.

Before moving on, we need to consider several other ape-man attributes that, if not exactly supernatural, might not appear fully credible. Lio sometimes describe ape-men as burying their dead, preferring to do so on or near beaches, where the sand doesn't require tools to dig, and as digging up relatives that humans have buried. If the ape-men were hominins (members of the genus Homo) this could be plausible. But more significant is a Lio belief, also found in other parts of Flores, that monkeys, too, bury their dead.

Heard more often is the claim that ape-men are afraid of a type of tall, stiff, hard-stemmed reed named *gai kiu*—possibly broom grass (*Thysanolaena maxima*). Accordingly, the reed can serve as a magical bane, to ward off ape-men or drive them away, for example from cultivated fields or livestock pens, where they might come to steal. Lio further say that by striking an ape-man with *gai kiu*, a person can seriously harm or kill the creature (as in a myth we'll review in the next chapter). Mentioned less often, but possessing similar efficacy, are a type of thin-stemmed bamboo (*Schizostachyum blumei*) and a stout grass called "dog's tail" (*éko lako*), either a dialectal name for *gai kiu* or a term referring to wild sugar cane (*Saccharum spontaneum*). The idea that ape-men are averse to a plant called "dog's tail" probably reflects the ape-man's well-known fear of dogs. And it may be no coincidence, either, that *gai kiu* stems are used for making arrows.[3]

Not just the hominoids but familiar animals are equally susceptible to certain banes. For example, to ward off monkeys, pigs, and other garden pests, cultivators use the bones and other body parts of dugongs (or seacows). Sometimes Lio contrast the susceptibility of ape-men to *gai kiu* and other grasses with their supposed invulnerability to metal weapons. For vine mothers, by contrast, metal is a bane. The idea may have its source in the exceptional physical strength of ape-men. All the same, it is clearly contradicted by reports of men shooting and killing hominoids with metal-tipped arrows. Also, such invulnerability is not exclusive to ape-men, for some Lio men claim to be equally invulnerable to weapons.

Revealed in Arndt's 1933 dictionary, another curiosity of ape-men is an inclination to prod or tickle people. While neither flying, disappearing, nor transforming receive any mention in Arndt's sample sentences, as many as three contain references to this habit—in one instance, in a sort of tickling competition, which the hominoids are described as always winning. Lio, however, rarely mentioned the habit to me. Since it would involve close contact between ape-men and humans, it is unclear in what circumstances tickling or prodding could occur. And as pointed out earlier in reference to Arndt's sample sentences, it's likely that the behavior is mostly an artifact of legends or some other sort of traditional story. Still, explaining how ape-men usually just "bother people" rather than seriously harming them, one young man, the previously introduced Mbaku, cited the example of unseen hominoids tickling or poking people from behind. The "victim" might then exclaim "who's that?"—to which the hominoid will echo, "who's that?" The person might turn around and see the tickler or nothing at all. According to Mbaku, this can happen when people are asleep, and they are awakened by the experience. But it may also occur around sunset when a person is sitting down, relaxing, and—though Mbaku didn't say so—we might reasonably add "dropping off." On two occasions I also heard that ape-men sometimes enter dwellings at night and sleep with people. In either case, it is very likely that the reputed experiences reflect dreams. A supposed eyewitness account I discuss in chapter 6 involved tickling ape-men—the only account to do so—and by all indications this concerned a dream remembered as a waking experience.

No indigenous Florenese spirits I've heard about tickle people. Western Indonesian sprites or elves, figures introduced to Flores during the last several decades and known only by the Javanese and national Indonesian language name "kurcaci," are sometimes described as ticklers. (We'll be meeting these figures again in a moment.) And so are similar beings in other parts of the world—including leprechauns. People elsewhere describe spirits as tickling people to death. Even so, lethal tickling is also attributed to obviously natural creatures. For example, Nigerian stories tell how a manatee (*Trichechus senegalensis*)—an Atlantic sea mammal—will, upon encountering humans, "tickle them so hard they drown."[4]

Supernatural Beliefs About Scientifically Documented Animals

It is already clear how Lio concepts of "flying," "disappearing," and "trans-forming" can accommodate things that do not, in fact, contravene natural laws (such as rocks first thought to be cows). It might still seem that ape-men share a good bit in common with entirely or largely supernatural beings. Yet in the Lio view this equally applies to several animals that Westerners would consider completely natural, for Lio describe these too as capable of changing shape or vanishing.

Animals Lio credit with powers similar to ape-men include dugongs (the species *Dugong dugon*), rarely-seen sea mammals of the Indian Ocean and southwest Pacific that, coincidentally, are close relatives to the aforementioned tickling manatees of the Atlantic. According to Lio, dugongs derive from a human being, an origin they also attribute to ape-men and monkeys. The first dugong was a woman driven into the sea by a cruel husband. Her lower body was then transformed into a large fish-like tail, a characteristic feature of *Dugong dugon*, while her upper half retained the nipples, or "breasts," of a mammal—or as Lio see it, specifically a human female. In light of these features it is no surprise that Western writers have recognized dugongs (and to a lesser extent manatees, which lack the fish-like tail) as the probable source of the image of mermaids or merfolk. With reference to their human origin, Lio call the sea mammals "dugong people" (*ata ruju*) or "dugong people fish" (*ika ata ruju*). For the same reason, they also deem the animals capable of transforming or reassuming human form, and in some contexts disappearing and making themselves visible to people selectively.[5]

As we'll see in a moment, dugong bodies, like ape-man bodies, are for Lio a source of supernaturally powerful relics. But how can we explain these sur-prising parallels? One reason appears to lie in a conception of both the homi-noids and the sea mammals as anomalously combining features of humans and nonhuman animals. Dugongs do this by combining the upper half of a human with the lower half of a fish. Ape-men, on the other hand, are erect and bipedal humanlike creatures that possess monkeylike faces and hairy bodies like animals. Lio may further understand dugongs and ape-men as complementary figures embodying contrasts of sea and land (or mountain forest), hairlessness and hairiness, and since Lio regard dugongs as typically well-disposed toward humans, benevolence and (partial) malevolence. From an external perspective, dugongs and ape-men are, respectively, scientifically documented and undocumented species. But Lio, of course, know nothing of this, as for them the two creatures are equally mysterious and equally powerful.

Other animals Lio credit with supernatural qualities include freshwater turtles, sometimes specified as "mountain turtles" (apparently *Cuora amboinensis*), and giant land crabs (the species *Birgus latro*, in English called coconut crabs). The world's largest terrestrial arthropod (a group including insects, arachnids, and crustaceans), these crabs are remarkable creatures, growing up to a meter (3 feet, 3 inches) in diameter from the end of one pincer to the opposite pincer. As noted, Lio believe both animals to be forms taken by forest spirits or, alternatively, possessions of the spirits. Either way, people claim the crabs and turtles can disappear and take the shape of humans. Encountering either sort of animal, therefore, causes great fear; they can harm people by invisible means, and killing them (which is strictly taboo) inevitably results in the deaths of offenders or their close relatives.

As with ape-men and dugongs, Lio characterize mountain turtles and coconut crabs as very rare and seldom seen. In fact, though both the turtles and crabs occur on other Indonesian islands, and though zoologists therefore consider their occurrence on Flores as credible and even probable, neither species has yet been scientifically documented for the island.[6] Although in a different way, the giant, highly terrestrial crabs and the mountain turtles are also ambiguous creatures. Lio, who are much more familiar with far larger sea turtles and far smaller species of land crabs, regard crabs and turtles of all sorts as animals that belong in or near the sea. Yet seemingly contradicting the natural order of things, mountain turtles and giant crabs, on the rare occasions when people see them, always appear well inland and often in highland forests. In a sense, dugongs, too, challenge the land-sea divide. Not only are these creatures, like ape-men, ambiguous in relation to the human-animal opposition; their bodies also incorporate features of both land creatures (specifically human beings in the Lio perception) and creatures of the sea (that is, "fish") even while they live exclusively in the sea.

My reason for comparing Lio ideas about ape-men to their ideas about these other animals should be clear. Fantastic powers comparable to those some Lio attribute to ape-men are identical to those they attribute to species recognized by science, or—especially in the case of dugongs—animals that Lio can definitely see. So supernatural qualities they ascribe to ape-men in no way reduce the possibility that the hominoids are equally natural species. As things rarely seen, ape-men and dugongs—as well as the giant crabs and mountain turtles—are, for Lio, creatures that are largely hidden, or "cryptic," and in this way, too, closely resemble unseen spirits. At the same time, it helps a lot that all four creatures are, in ways already discussed, visibly peculiar. And for this reason, Lio regularly describe encounters with ape-men, giant crabs, and freshwater turtles (but not dugongs) as alarming and threatening to personal well-being. So, in the same way, are encounters with spirits, beings described as invisible but which, nonetheless, occasionally manifest themselves to certain people in visible form.

Belonging to an invisible realm, spirits, in a sense, are things that should not be seen, even though seeing them in either animal or human form is supposed to benefit some people. Here I should mention that the Lio word for "see" (*tei*) further means "to encounter, meet with," "find," "contact," or "get, obtain (something)." The question of visibility reveals another similarity between ape-men, other rare animals (including dugongs), and spirits. Lio maintain that not just anyone can see these things, and that if a person went looking for any of them, they would never find one. Put another way, encountering an ape-man, like a rare animal or a spirit, is always "unexpected, coincidental" (Indonesian "kebelutan"), a matter of "chance" or "luck" ("rejeki") or, when benefits follow from the encounter, a sign of being favored. So Lio conceive of the ability to "see" (or "encounter") such things as a special personal quality of individual observers.

Showing ape-man encounters to be personal in a more positive way, some Lio claimed that only people who possess "clear eyes" (Lio *mata dega*) can encounter either ape-men or spirits (especially vine mothers). The phrase refers to an exceptional visual ability, typically associated with mystical practitioners, or "men of power." Yet by no means do all Lio agree that ape-man sightings require such special ability. More importantly, most people who said they'd seen the hominoids, including eyewitnesses reporting the sightings that seemed most credible, were not people whom others credited with having "clear eyes," nor did they themselves claim this quality.

Nowadays people suggest that their inability to see certain things may reflect general changes in local culture. Once, when discussing rare freshwater turtles, an elderly lady remarked that since people have become Christian, they no longer "see" or "encounter" (*tei*) the creatures. Interestingly, on a separate occasion, a younger man said exactly the same about ape-men. This may simply be a way of saying that ape-men were more numerous in former times. Or, since the Catholic Church was mentioned, it could suggest that one needs to believe in such things in order to see them. But there's nothing to show that people no longer regard ape-men as extant natural creatures. Nor does the Church have anything to say about the hominoids, let alone anything that might discourage people from "believing" they exist. So the statement more likely implies that modern people have difficulty seeing such things because they belong to an older way of life. In the case of the mountain turtles and giant crabs, there's the further factor of their association with decidedly non-Christian forest spirits. This doesn't apply to the hominoids. But their rarity and, we might also say, uncanniness put ape-men in the same category of mysterious things.

Ape-men, Imported Spirits, and Modern Media

As mentioned earlier, there's more to say about ape-men flying. So far as I'm aware neither dugongs nor any sort of turtle or crab is credited with this ability. Gliding gaits aside, some Lio apparently understand flying hominoids as defying the law of gravity and flying or gliding through the air like a bird or plane. Actually, though, there's good reason to believe that this idea is recent.

Once I was talking to Maga, a man in his forties who occasionally served as a field assistant, when the subject turned to fantastic powers sometimes attributed to ape-men. Most of what he knew about the hominoids Maga said he'd heard as a youngster from his parents and grandparents. And he was quite certain he'd never heard his elders describe ape-men as able to fly or transform. He therefore concluded that these powers were not authentic parts of any traditional knowledge of the creatures. Maga was one of the six Lio who denied ape-men could change their shape. Interestingly, four of the others were over seventy.

So where did the idea that ape-men could fly and transform come from? The short answer is from supernatural figures "imported" from western Indonesia, and mainly from Java. In recent decades Flores Islanders have become familiar with these figures from modern media images—and since the 1980s or 1990s, primarily from television. For the most part, I refer to elves or sprites that Lio and other Florenese call by the Indonesian and Javanese name "kurcaci."

From what I learned in several parts of Flores, these kurcaci first became known in larger towns sometime after national independence in 1945, but in rural locations probably not before the 1960s or 1970s. In the Nagé region of west central Flores, stories of incidents supposedly involving kurcaci, notably thefts, date from the 1980s. While I was staying in Nagé in 1984, a young boy reported seeing one or more "kurcaci" inside his village in the middle of the night. My inquiries suggested he had learned of the creatures from illustrated children's books of Javanese origin available in school collections—at the time one of the few sources of reading material or modern media of any kind on Flores. Nagé people, it is important to note, identify kurcaci neither with reputedly extinct local hominoids—creatures named "ebu gogo"—nor with indigenous forest spirits. By contrast, Lio have come to employ "kurcaci" synonymously with "lai ho'a," the name for their local ape-men.

During the last three decades, another Javanese supernatural has entered the imagination of Flores Islanders. This is a sort of magically powerful elf, in Indonesian called "tuyul."[7] In spite of the distinct name, Lio and other islanders usually refer to these elves, too, as "kurcaci"—thus using this name (apparently because of its historical priority) as a cover term for more than one variety of imported spirits. Since the 1990s, the elves have frequently appeared in Indonesian television programs or Javanese and other Southeast Asian films shown on television. One of the most influential has been a popular weekly program called "Tuyul and Mbak Yul," the second name belonging to the main human character, a rather fetching young woman. Still rerun occasionally, the program has been popular among Lio and other Florenese since the 1990s, when it was first aired. In this, actors playing tuyul included an unusually short adult and several achondroplastic dwarfs. By all indications the majority of Lio first became familiar with the Indonesian word "tuyul" through exposure to this show.

As the small size of these tuyul should suggest, the TV elves resemble the Lio ape-men in several respects. Not only are they portrayed as short-bodied hominoids, they are considered facially unattractive, and in Javanese folklore—and now pan-Indonesian popular culture—tuyul (or kurcaci) are depicted as thieves. True, the mischievous elves use magical means to steal, including the power to become invisible, whereas Lio describe ape-men as straightforwardly taking crops from fields like any natural animal. But Lio also accuse ape-men of taking food from dwellings and small livestock from settlements, and when they do, typically under cover of night, they too are normally unseen. In addition, stories of kurcaci I recorded outside of Lio (including in the Nagé region) depicted the tiny creatures, like the Lio hominoids, as being exceptionally strong.

So what has all this to do with flying? Actually, quite a lot. For in media depictions, the wingless tuyul (or kurcaci) fly or glide through the air—rather like Superman—change shape, and disappear and reappear, usually in another location. Explaining how ape-men "fly," Tangi, a young man mentioned earlier, explained that, lacking wings, the hominoids don't fly like birds but glide through the air, perhaps 5 meters (over 16 feet) above the ground. Tangi also claimed that only people with supernatural power (probably meaning "clear eyes") could actually see this. In another case, Nuwa, an elderly man with whom I discussed ape-men several times, compared the hominoids to small-statured figures he'd seen on television. Interestingly, Nuwa never mentioned flying in his general account of ape-men, which in fact comprised one of the most naturalistic descriptions of the hominoids I ever recorded. Yet further questioning about the media figures revealed that he was referring to the elves called "tuyul." What's more, it turned out that the old highlander, being unaware of the possibilities of modern technology, thought the TV elves were real creatures, and in this respect it seems unlikely that Nuwa, a former administrative headman and respected local leader, was unique.

The conclusion is therefore clear. Other explanations can be found for seemingly supernatural powers sometimes attributed to ape-men—in details of Lio language, culturally informed experiences of other rare and elusive animals, and so on. Yet a major source of these ideas in recent years, especially the notion that ape-men can fly, is found in regular exposure to moving images of physically similar and apparently substantial beings exercising the same powers in modern media. It was not just old Nuwa who identified the exotic elves with local ape-men. Nowadays, Lio use "kurcaci" as the Indonesian translation of "lai ho'a" (ape-man). This might be taken to suggest that Lio identify local ape-men, too, as primarily supernatural beings. However, it's important to stress that Lio never apply "kurcaci" to any local spirits—including forest spirits and vine mothers—a point I took pains to confirm. Evidently the reason is that, as portrayed in film and formerly in illustrated books, the imported beings, unlike incorporeal local spirits, always have bodies, and possible shape-shifting aside, bodies that normally take the form of a small hominoid. The Javanese elves are also characterized as stealing material objects, something that Lio spirits never do. And especially when identified as "tuyul," the elves can be captured or purchased like slaves, so that avaricious owners can instruct them to steal on their behalf. As we'll see in the next chapter, Lio describe ape-men, too, as sometimes being caught—a susceptibility that, of course, doesn't apply to bodiless spirits.

All the same, not all Lio equate kurcaci and ape-men, or do not do so consistently. Some older people distinguished them, remarking for example that what I'd heard others say about indigenous hominoids actually applied instead to the newer kurcaci. Indeed, the two are by no means identical. For example, the television elves are not hairy-bodied and they wear clothes. In addition, they are usually depicted as occupying modern urban settings, whereas ape-men, of course, are creatures of highland forests. Nevertheless, when speaking about "kurcaci" I sometimes found it difficult to know whether people were referring to ape-men or something else—for example, when they spoke of kurcaci entering human settlements (as in Mbapu's story, mentioned above). As "kurcaci" are commonly described as just 50–60 centimeters (20–24 inches or 2 feet) tall—and as one man insisted, never taller than this—it's also possible that a conflation of ape-men and exotic elves has, as it were, reduced local estimates of the hominoids' size.

Besides certain physical and behavioral similarities, there's another reason why Lio would want to identify ape-men with Javanese elves. Partly because "kurcaci" and "tuyul" are terms from the national Indonesian language, and because visual images of the elves come from a very centralized national media, Lio are aware that these beings ultimately belong to a national culture, which they implicitly consider superior to their own. For this reason alone, inhabitants of marginal areas like the eastern Indonesian islands are motivated to identify elements of local culture with the largely western Indonesian (and mostly Javanese) national culture. And, by the same token, Lio should be inclined to view the media elves, in spite of—or because of—any recognized differences, as truer to the type or more accurately depicted than partly comparable local hominoids.

Ape-man Bones and Other Relics

As noted, a practice linking ape-men with dugongs is the Lio use of body parts of both to obtain magical or supernatural benefits. Taking a cue from values attached to corporeal remains and other items associated with Catholic saints and other religious personages, I collectively call these "relics." Curiously enough, people in other regions of Flores seem not to employ body parts of locally recognized hominoids as relics, so the practice appears unique to Lio.

Dugong relics include the creatures' ribs and other bones, hide, oil, and teeth, all of course retrieved from the carcasses of dead specimens (see figure 3.1). But most valued of all are dugong tears, which Lio say are efficacious only when "given" to a person by a living animal. Ape-man relics are similar. These include bones, teeth, and nails. Less often, people mentioned hair, spittle, feces, blood, testicles, and eyeballs (the last three only being mentioned once). However, the relic Lio refer to most often is ape-man urine, kept like dugong's tears in tiny bottles. Because urine comes from the lower part of a creature's body and tears come from the uppermost part, readers might recognize a symbolic contrast between the two substances—a contrast that further fits the entirely benign image of dugongs (the most maternal of animals, according to Lio) and the partly malign and fearsome character of ape-men.

When carried on the person, Lio say ape-man relics protect possessors, either by rendering them invulnerable to attack or by making them invisible. They also enable people to produce abundant harvests, as explained in legends reviewed in the following chapter. Somba, a ritual specialist living on Lio's south coast, described how a parent could dip an ape-man bone in coconut oil before rubbing the oil on the body of a baby boy. The infant will then grow up to be invulnerable "just like an ape-man," and will be able to steal things without getting caught. Lio more generally mention achieving invisibility simply by carrying hominoid relics on their person. Such ideas reflect a universal principle anthropologists call "sympathetic magic," whereby "like affects like." In this case, ape-men's great strength and their reputed invulnerability to metal weapons are transferred by way of their body parts to human possessors. Similarly, if one places ape-man urine on a bush-knife or bullet, the object will always hit its mark.

Like the previously mentioned notion that only people with "clear eyes" can see ape-men, invisibility is evidently linked with the hominoids' ability to "disappear." In regard to the hominoids' general resemblance to monkeys, however, a more obvious factor is a belief that a person can equally achieve invisibility by obtaining a particular bone of a monkey, taken from a skeleton and identified through trial and error. In the same way, a person can gain protection from the bones of humans believed to have possessed exceptional powers during their lifetimes. According to the aforementioned Somba, a deceased fellow villager once disinterred the bones of his father's cousin, a reputed witch, and took one of these with him when he traveled to Malaysia.

In a more charitable vein Lio healers may employ ape-man relics to cure people, for example, by rubbing a bottle of urine on a patient's body or by applying what possessors claim to be an ape-man's spittle. The same use is made of dugong relics. But ape-man relics are much rarer than dugong relics. People say ape-man body parts were more numerous in former times. They also mention these being sold or "given away" to outsiders—including Indonesian troops sent to the Flores after the destructive tsunami of December 1992. In some instances, moreover, people reputed to own ape-man relics actually did not, or no longer had them in their possession, or were reluctant to show them to me—except, in some instances, in exchange for ridiculously large sums of money!

Despite these difficulties, I did get to see some relics. But I must now disappoint readers, for nothing I saw looked like the remains of an unidentified hominoid. There were primate remains to be sure. Yet the one I was able to have professionally inspected, by the late Colin Groves (a primatologist and polymathic anthropologist at the Australian National University), turned out to be a skull fragment from a monkey, a long-tailed macaque. In a similar vein, a photo of what was claimed to be an ape-man skull (owned by someone on Flores's north coast whom I was never able to identify) was clearly the cranium of a young orangutan. At least the monkey and orangutan skulls conform to the usual representation of ape-men as monkey- or ape-like creatures. But most hominoid relics I saw didn't look anything like the remains of primates. One skull was evidently a puppy dog's; a collection of bones and teeth came from a rodent, possibly the Flores giant rat (*Papagomys armandvillei*); while two teeth appeared to belong to a dog and an ungulate. Other bits of bone I was shown, including a 6-centimeter (2.4-inch) fragment the owner claimed had come from an ape-man's forearm, I was unable to identify.

In just one instance putative relics were artifacts. Pora, a retired school-teacher but also widely known locally as a grifter, showed me a small knife and a metal pendant depicting a humanlike figure that he claimed had belonged to an ape-man, together with a small bottle he said contained ape-man urine. The manufactured objects, of course, fly in the face of the Lio view of the hominoids as lacking any material culture. Further inquiries revealed that the pendant was a standard item of the sort Lio use as bride price, and that the figure represented an ancestor, not an ape-man. People delivered the same verdict on a roughly-carved human head on the detachable wooden hilt of the knife. The knife, I learned, was a type Lio use in rice-harvesting rituals—and such knives must always have a human figure on the hilt, because according to a myth known throughout Lio country, rice originally derived from the body of a human female.

Two other men showed me hard round objects, both evidently of modern manufacture, that they said derived from ape-men. Gali, the owner of the aforementioned piece of a monkey's skull, brought out a heavy metal ball the size of a pea, which he said he'd found inside an "ape-man's head." The other man possessed a similar ball. But his account of how he'd obtained this was even less clear.

As we'll see from stories reviewed in the next chapter, Lio depict people who in former times obtained ape-man relics from hominoids caught in traps or snares set to catch other animals. The men usually receive the relics in return for the creature's release. If not buried with human recipients, such relics are passed on to their children and descendants. Pointing to a strong personal connection between the items and their original owners, Lio say ape-men relics should never be sold or otherwise alienated—only temporarily "loaned" to other people. But nowadays, it seems, this injunction is not always followed. I met a number of people who clearly hoped I would purchase such items—typically for exorbitant sums (in one case equivalent to several thousand dollars).

Though short on details, a few older stories concerned deceased men who'd obtained bones or teeth from the bodies of ape-men shot with arrows or killed by dogs. In more recent times, as well, men have obtained relics from the carcasses of what they or others interpreted as hominoids. In 2014, in central Lio, I met an unassuming man in his sixties named Reto who related how, the previous year, he'd had a dream in which a beautiful woman appeared, telling him to dig inside a certain small cave. Though he wasn't sure, he thought the woman might have been an ancestor, and since she'd told him to excavate, Reto hoped to find something made of precious metal. As it happened, he dug up the skeleton of a large rodent, the probable *Papagomys* rat mentioned above (see figure 3.2).

Similar to Reto, Fara—a seemingly depressed though not unforthcoming younger man—let me inspect bones he'd found some years previously near a spot where he had set a snare for junglefowl. Before showing them he described the remains as looking like the bones of a human or a dog. In fact, one was the puppy dog's skull referred to earlier. On the following night, Fara had a dream in which a woman appeared—whether she was beautiful he didn't say—identifying the remains as her own. When I asked him, Fara could not (or would not) confirm whether his relics came from an ape-man. Yet fellow villagers I'd spoken to before our meeting were sure this was precisely their origin. With Reto, too, the idea that his rat bones were hominoid relics had apparently originated with other people. In view of the actual appearance of these bones, as well as the skull of the young dog, it is an interesting question whether such interpretations are connected—as cause or effect—with the notion that ape-men can transform into other animals.

Besides bones, a couple of men spoke of "stones" derived from ape-men, but I never got to see any of these. It is important to note that putative ape-man relics are by no means the most common objects Lio deploy to magical ends. As noted, dugong relics are more common. So, too, are stones and other things obtained from forest spirits and ancestors. Because most of these stones come from the bodies of snakes and eels, I hesitate to call them "relics" as their ultimate source is not an animal but a bodiless spirit. Anyway, because these spirit stones equally provide protective and healing powers, relics claimed to come from hominoids are by no means essential to Lio mystical beliefs and magico-religious practice. As mentioned, the ape-men themselves are in no way central to Lio culture. In fact, by contrast to ancestors, witches, and forest spirits, they are decidedly peripheral.

It is already clear that ape-men, though sometimes credited with supernatural power, have little in common with pure spirits (for example, forest spirits). They are far more like unusual humans (witches and "men of power") and unusual animals. Of course, none of this "proves" that ape-men exist in the same way as humans or other scientifically documented species. But what the comparisons do show is that, for Lio people, the hominoids are no more supernatural than some of these, notably dugongs—and therefore no less natural. As we'll see, an investigation of local myth and legend leads to the same conclusion.

PART II

STORIES OF
APE-MEN

4

APE-MEN IN MYTH AND LEGEND

Anthropologists use "myth" to distinguish stories set in a timeless past that typically concern spiritual beings or human or animal characters with supernatural powers. Indeed, so miraculous are things told in myths that even their narrators may consider them improbable or impossible, the more so if the events were described as occurring in the present day. Whether people fully "believe" in their myths is often difficult to tell. Whenever I asked Flores Islanders whether something described in a myth "actually happened," they often shrugged and said they did not know. This is not to say that myths are not valued or don't convey any sort of truth. Nor does it mean myths have no connection to historical events. But it does mean that people don't always accept everything in their mythology as an accurate reflection of the way things actually are. To a lesser extent the same applies to legends, often distinguished from myths as stories concerning known historical individuals.

The Origin of Ape-men

Lio seem to have no widely known or generally recognized myth about the origin of ape-men. Stories describing the derivation of many familiar animals are also rare. Still, a few people had something to say about how the hominoids originated.

Some people simply described ape-men as deriving from human beings. A few suggested they developed hairy bodies because, unlike ordinary people, they live an animal-like existence in forests. According to one story, told by a man who'd heard it from a native catechist, ape-men were originally humans who were "driven out" of their native communities by God. This was because, being stronger and more aggressive than their fellows, the hominoids' ancestors "killed people, stole, and did not marry properly." As a result, ape-men cannot live permanently anywhere. If they try to take up residence in a cave or tree, something will soon drive them away. Alluding to the idea that ape-men steal hens' eggs, the narrator explained that ape-men offer eggs to God in an attempt to regain a permanent place in the world. But God will never accept their offerings.

The parallels with Christian notions of wandering Jews and the rejection of Cain's offering by God, who cursed Cain to wander the earth, are rather obvious. So in all probability the story has a biblical and therefore a twentieth-century origin. Even so, the idea of ape-men lacking a permanent home is reminiscent of the earlier-mentioned notion that the creatures do not live permanently in given locations but move constantly from place to place. Ape-men being stronger than their original human companions corresponds to a more general idea. Notions of their ancestors killing people and marrying irregularly do not. Nevertheless, Lio compare irregular marriage to the mating of animals, which as they remark, is not rule-governed like human marriage. So this attribute implicitly identifies ape-men as animals rather than people.

The theme of irregular sexual connection appears again in the only other account I heard of the origin of ape-men. The narrator was Goda (see figure 4.1), a Lio elder we've met before. The myth appears to be modeled on a much better known story concerning the origin of monkeys. Yet Goda, a man in his seventies, insisted he had heard it from his long-deceased uncle. Here is a summary:

An elderly brother and sister became involved in a sexual relationship. The sister was long past childbearing age. One night a younger
relative spied on the pair, saw them engaging in intercourse, and
told others about this. Following their transgression a terrific storm
arose, with thunder and lightning. Fellow villagers then realized
that, to stop the storm, they would have to formally banish the
transgressors. As part of the procedure, the villagers exclaimed,
"you will become a kind of animal, what kind of animal is up to
you." Their kinsmen further declared, "you are no longer our relatives, you will become lai ho'a (ape-men)." They then chased the
elderly siblings out of the village and into the forest, where they
mixed with monkeys.

Goda related this tale after he'd described how monkeys, too, derive
from humans. As he further explained, while ape-men and monkeys have a
common origin, or derive from a "single stock," it was monkeys that "split
off" first. Specifying the mythical couple as elderly obviously agrees with the
more general idea that ape-men physically resemble old people. And as Goda
further remarked, the fact that they were past childbearing age explains why
ape-men are not prolific and their numbers are low. By contrast, a version of the
myth recounting the origin of monkeys features an adolescent brother-sister
pair—aged about fourteen or fifteen, Goda reckoned. They, too, committed
incest and were banished into the forest by their parents.

Cave-burning Stories

Also located in an indefinite past, another Lio myth describes humans
killing ape-men by setting fire to caves with the hominoids inside. Interestingly, we find the same theme in several parts of Flores Island as well
as much further afield, in Sri Lanka. In Lio country I recorded half a
dozen versions, all associated with different locations. Recorded in 2014,

the following version, told by a middle-aged man named Ruju, is the longest and most detailed:

> Long ago, people of Tana Langi village were holding a feast. They performed a circle-dance. A number of ape-men also attended and joined in the dancing, but they did so in disguise. Before arriving they changed into humans. A village woman wanted to join the dance but could not, because she was cradling her baby. Noticing this, one ape-man changed form so that he looked just like the woman's husband, and he offered to hold the baby while she danced. The woman accepted, but while she was dancing the hominoid ran away with the infant to the ape-men's cave. Shortly before this, the other hominoids had returned to their cave, taking with them gourd eating and drinking vessels that the villagers had used while feasting. Discovering the child missing, the villagers mounted a search and eventually arrived at the cave. The ape-men agreed to surrender both the abducted infant and the gourd vessels, in exchange for a large quantity of palm fiber. They wanted to use the fiber as sleeping mats. So the villagers collected a large amount of fiber and tossed it inside the cave. But when they were sure all the ape-men were inside, they set light to the fiber. Fleeing from the smoke and flames, the hominoids tried to escape by way of an exit hole, located at the back of the cave. As they escaped, people tried to kill them with bush-knives and clubs. But the ape-men proved invulnerable to these weapons. The villagers then tried hitting the creatures with lengths of *gai kiu* (the bamboo-like reed discussed in the previous chapter), and thus were able to strike them dead. However, the ape-men left two infants of their own, of opposite sex, behind in the cave. The two managed to escape by exiting another way, and so it is that ape-men survive to the present day.

Other versions of the same myth are less detailed. Concerning the now deserted village of Mbira Mbo'a, which according to one report once bore the name Lai Ho'a (Ape-man) or Lia Lai Ho'a (Ape-man Cave), one story describes how a small child was playing in the village when an ape-man came along and snatched it away. The villagers eventually found the cave where the ape-men were holding the kidnapped infant, and they managed to wrestle the child from a female hominoid that was cradling it. To prevent further abductions, the villagers set the cave alight with palm fiber. The conflagration killed five ape-men. But because the villagers feared that one or more of the creatures may have escaped, they decided to abandon Mbira Mbo'a and established several new villages.

Like myths the world over, such stories sometimes serve political or historical ends, in this instance rationalizing the movement of people and the founding of settlements—and also legitimizing the occupation of current territories. One of these settlements was Tana Langi, the setting of the more complex story of ape-man destruction related above. But more important for our purposes, the Mbira Mbo'a story contains none of the fantastic elements included in the first myth. Specifically, there is no mention of the ape-men transforming into humans or being susceptible to the plant *gai kiu*.

In fact, the same goes for other Lio cave-burning stories I heard, none of which mentions transformation or any comparable supernatural action. In one, ape-men abduct a human child and leave one of their own in its place. This theme of the changeling occurs in myths about similar hominoids told in other parts of Flores. Indeed, it appears in many other parts of the world, where the culprits are not ape-men but familiar animals, humans, witches, or spirits. Lio never mentioned abduction of children (or adults, for that matter) as a hominoid habit in general descriptions of ape-men. By contrast, people often describe ape-men as stealing crops and small livestock, and this more likely possibility figures as the motivation for cave-burning and extermination in all other Lio stories in this genre.

Crop theft also motivates cave-burning in a myth from west central Flores. Here, people of the Nagé village of 'Ua relate how, long ago, their ancestors conspired to exterminate another group of hominoids, the previously mentioned ebu gogo. Nagé describe this act as resulting in the complete extinction of these creatures, at least within Nagé territory. But a more interesting comparison lies in another detail. For, like the ape-men exterminated by the people of Tana Langi, the ebu gogo took the villagers' gourd vessels—though in this case they swallowed them, thinking they were edible. The Nagé hominoids also joined in circle-dancing, though they did not transform into humans in order to do so. And as in the Tana Langi myth, two of the creatures—also a brother-sister pair—managed to escape the conflagration, by fleeing to a location far away from Nagé territory.

If swallowing gourd plates seems fantastic, it is the only such event included in the Nagé story, which like the majority of Lio variants makes no reference to shape-shifting or similarly fantastic acts. Also, as in most Lio versions, it is simply the fire and smoke that kill the hominoids, not some magical bane. But despite these differences, all the myths are, apparently, variants of a single, quite specific Flores-wide tradition.[1] Present in all versions is the act of setting fire to caves using palm fiber, specifically fiber from the trunk of the Arenga palm, a highly flammable material when dry. Cave-burning, we should note, is an efficient and very practical way of exterminating pestilential cave-dwelling creatures that pose a threat to human livelihoods—including reputedly very strong hominoid crop and livestock thieves that villagers couldn't easily deal with in any other way. In fact, Flores Islanders also employ burning to kill or drive away feral cats that inhabit rock crevices and prey on chickens.

In contrast to these naturalistic details, ape-man transformation and the creatures' vulnerability to the plant called *gai kiu* are mentioned only in the Tana Langi variant, recounted above. These single instances, therefore, suggest an elaboration of a more basic tradition. And since I recorded this version in 2014, it's not impossible that transformation, especially, reflects influence from the foreign media images described in the last chapter. Alternatively, the fact that one of the Tana Langi ape-men assumes the form of a specific person could reflect a power that Lio normally attribute to the spirits called "vine mothers."

So realistic is cave-burning as an efficient way of killing troglodytes, it's not surprising that such an event was once interpreted as occurring not in a timeless past but very recently. In 2003, Iku, a Sikkanese man who'd moved to northern Lio to cultivate land, told me a story he'd heard from several young Lio men who once worked for him harvesting rice. The story concerned Lio hunters who, apparently just a few years previously, had chased an ape-man into a cave and set the cave alight. Afterward, they discovered there had been another hominoid inside, which was also burned to death. According to a different version the same man related in 2014 (thus eleven years later) the hunters set fire to a cave in which they thought a wild pig had taken refuge, and they subsequently found three or four dead hominoids inside. But this inconsistency hardly matters. In the same year I was finally able to locate one of the Lio men who'd told the story to Iku. And he claimed that the events actually occurred long ago and, moreover, formed part of a myth similar to the one involving the Tana Langi circle-dancers.

In view of the distinct scenario of hunters chasing either ape-men or pigs, this connection might seem unlikely. Still, it's possible, even probable, that not just Sikkanese migrants but Lio themselves sometimes "relocate" mythical events in a far more recent past, and in the context of more mundane activities. Another possible example is a story set in the early twentieth century concerning a cultivator who drove a number of hominoids from a cave not by burning but with a modern device called a "bamboo cannon" (a hollow length of thick bamboo into which a quantity of kerosene is poured and ignited to create an explosion). But if people do still use such methods to kill or chase away creatures taken to be ape-men, I never met anyone who claimed to have witnessed such an act.

Tales of Capture

Folk traditions of mystery hominoids often include stories about people capturing them.[2] In fact, capture is implicit in Florenese cave-burning stories insofar as these always involve people, in effect, confining the creatures in an enclosed space. By contrast, other Lio stories of ape-man capture concern thieving hominoids getting accidentally caught in traps set for pestilential wild animals inside or near cultivated plots. Narrators most often mentioned wild pigs as the intended victims, but a few specified other animals, including dogs, rats, civets, and monkeys.

Typically, the garden owners subsequently find the trapped hominoid and, in return for the creature's release, obtain its body parts or products. The captor then keeps these as magical relics, as described in the previous chapter. Since in all accounts I recorded, the beneficiary was a named historical figure—usually a relative of the narrator who lived during the twentieth century—such stories might be distinguished as "legends" rather than "myths." Even so, the stories usually include fantastic elements of the sort more characteristic of myths. Though some of the events supposedly took place just a few decades previously, the human character in these tales was always someone already dead. Also, it was never clear whether the narrator had heard the story directly from the person reputedly involved, and in many instances the story appeared to reflect a more general knowledge. So it's quite possible that some capture tales describe events that captors would have denied while still alive, and that the legends have mostly grown up around renowned people after their deaths.

I recorded seventeen such stories. Although somewhat more fantastical than others, a good example concerns a man named Leja who was probably born in the first decades of the twentieth century and may have died as recently as 2008. A man whose name and exploits are widely known, Leja has become something of a culture hero in Lio country. Told by his cousin's son, Koba, this is his story:

> Once Leja set a snare for wild pigs in a forest somewhere on the mountain named Ndura. A female ape-man got caught in the snare and asked to be released, but Leja refused, saying he was afraid. The creature said there was no need to be afraid, it had come only to do him good. So Leja agreed. The hominoid then offered Leja

three stones of different colors. But he did not want these. He only wanted the ape-man's urine. As Leja had no container, the creature reached behind its back and produced a small bottle in which Leja could collect the urine. He then released the creature. After that they became friends, and the female ape-man accompanied Leja wherever he went.

In subsequent conversation, Koba told me that only Leja was able to see the ape-man and only he was able to hear it when it spoke. No other capture tale included these details, though in another story other people were able to see a captured ape-man only when the creature and the observer were both close to the captor.

Other stories in this genre depict ape-men communicating with their captors, but usually without saying how they did so. Leja's story included even more unusual details. According to Koba, both the female ape-man and her hominoid father had personal names, and she was "simple-minded." The female's name was Ere and her father was called Re'e. Appropriately enough, in view of usual descriptions of ape-men, Re'e means "ugly, bad" while Ere looks like the same name with the two syllables reversed (or what linguists call "metathesis"). What might be the significance of the female hominoid being simple-minded, I'm unable to say. Nonetheless, the attribution seems like a rather prosaic detail and, as such, offsets more fantastic elements of the story. (I recorded just one other mention of ape-men having personal names, but this was revealed to someone in a dream.)

Among the powers Leja gained from obtaining the ape-man's urine, he became able to complete work in his garden extremely quickly and with little effort. To ensure bountiful harvests, however, Leja was prohibited from bringing the urine into his dwelling or granary. When he was not carrying the urine on his person, he had to hide it in a secluded spot in the forest. Otherwise, produce kept in his hut or granary would quickly become depleted "because every time Leja took a portion, the ape-man would [invisibly] take an equal quantity for itself." A similar taboo observed by men enjoying this sort of benefit concerns taking food from granaries; they should always have someone else do this, otherwise the stores will, for the same reason, soon run low.

Responding to my question, Koba described Leja as possessing other extraordinary attributes, including the ability to fly and to disappear and reappear. These powers, he confirmed, Leja also obtained from his captured ape-man. More remarkably, though, the narrator attributed the same powers not only to Leja but also to a European Catholic priest, identified only as "Tua To" (Mr. To, possibly the German surname Toch), who was stationed on Flores from the 1940s to 1960s. A confirmed pagan, Leja supposedly once engaged in battle with the priest, himself something of a legend who, like other European priests at this time, Lio credited with several supernatural powers. The battle ended in a draw. But more significant for our purposes, the details of the clash—including each contestant flying through the air, disappearing and reappearing, and seemingly dividing into several bodies—sounded for all the world like something Lio might see in Indonesian or Chinese films that, in recent decades, have regularly appeared on TV.

In just one other instance did a man obtain the ability to "fly" from hominoid relics, specifically an ape-man's tooth. Like Leja, people who gained favor from captive ape-men were reputedly able to work land quickly and easily and produce abundant harvests. On the whole, though, details of stories about the accidental capture of ape-men are more varied. Other benefits include powers of invulnerability and invisibility, as well as healing power. Where narrators specified the body part or substance obtained from a captured ape-man, urine (mentioned in 6 of 12 stories) was the most common; other items included bones (in two instances), teeth, tears, feces, spittle, and toenails. Capture narratives usually don't specify how living ape-men were able to transfer urine or body products like bone to their captors. Presumably, though, they do so magically, as Leja's hominoid did when it suddenly produced a bottle from behind its back.

Another man spoke of a captor gaining power from an ape-man by "exchanging bodies" with the hominoid, without a transfer of particular physical relics. What exactly this involved I was unable to clarify, but a more straightforward exception is a single instance where a captive hominoid surrendered some of its spittle. Lio healers in general—including those who do not claim to possess ape-man relics—commonly administer cures by applying their spittle to the patient. In this instance, the ape-man spat into the captor's palm. Then, years later and by the same means, the man transferred this acquired power to his nephew, who told me his uncle's tale. The same story included an even more curious detail. Before accepting the spittle, and hence the ability to heal, the captor asked the ape-man to give him the power to come back to life three days after his death, a request obviously reflecting Christian influence. But the hominoid, evidently another speaking ape-man, responded that this was not possible as "we too live and die," meaning that death is inevitable for both humans and ape-men. The exchange is instructive, as in spite of the fanciful details typical of capture tales, it reveals, once again, a representation of ape-men as mortal creatures, quite different from immortal spirits.

A few stories describe an ape-man's capture and release initiating a continuing relationship with the victim. The tale of Leja is one example. In another story, an ape-man again visited the captor the evening after its release, and later its offspring also visited him from time to time.

Most stories of accidental capture don't mention the sex of the captive; where they did, one was male, two were female, and another was a male-female pair. Where the captive was female, only once did a narrator describe the captor as "marrying" the hominoid. But the son of the deceased captor later denied this—not surprisingly, perhaps—and the narrator added that the couple never had children! With Leja's female captive, there was no suggestion that he had "married" or become sexually involved with the creature.

In a few stories, events depart significantly from Leja's tale. Three concerned captives that were simply released. In one, a man named Kéli heard something on the roof of his field-hut in the middle of the night. He reached out and seized it, and discovered it was an ape-man. After speaking to the creature—what Kéli said the narrator did not know—he simply let it go. Kéli therefore obtained no relics or any sort of supernatural benefit, nor did people in the other two cases where ape-men were set free. When I asked how Kéli was able to capture an ape-man with his bare hands, the narrator immediately responded that this was possible only because, at the time, he had been naked. Later, a healer named Tipa (a man we've met before and will meet again in chapter 9) claimed that people only ever meet ape-men when they are naked. All sighting reports contradict this. But interestingly enough, Lio also say that men wanting to attract dugongs (sea-cows) to obtain their supernaturally powerful tears must remove all their clothing before entering the sea. Similarly, in the Nagé region I was told that a person had to be naked to catch a kurcaci, the previously described elves "imported" to Flores from western Indonesia several decades ago.

According to another story describing simple release—in fact the first tale of ape-man capture I recorded, in 2005—a group of villagers in northern Lio once found an ape-man hanging upside down in a snare set for wild pigs and "weeping like a small child." People who'd never seen such a thing before fled in fear. But people who knew what it was performed a ritual to protect themselves from possible consequences of the untoward event, and then set the creature free.

In two other stories, entrapment kills the hominoid victims. In one, an ape-man died in a kind of baited trap used to kill monkeys. In the other, a maize-stealing hominoid was caught in a pig snare and released after the garden owner warned it never to return. The creature, however, did come back, and it got trapped again. Caught in the snare, it either died or the owner killed it (which was not made clear). Eventually, the man obtained relics from the victim by burying the body and later disinterring its skeleton.

In some ways similar, another story concerned Rindo, an elderly bachelor who died just a year before I heard his tale. Surviving relatives told how, as a young man living alone in a field-hut, Rindo was visited by a young male ape-man and its mother. As the hominoids had stolen food from the hut, he warned them not to return or he would do them harm. But they did come back, so Rindo cut off one of the creatures' hands. Rindo kept the hand, as it held special powers, including the power to cure. The story did not reveal whether the victim survived the amputation. However, like the man who buried and later disinterred a dead ape-man, in this case it's obvious how someone was able to obtain a relic from an ape-man's body.

Just one story describes an ape-man escaping from a trap. After a cultivator discovered a specimen caught in a snare, he tried to strike it with a bush-knife. But the creature struggled, the snare broke, and the ape-man fled. Two days later the man fell ill and died—apparently either because he'd tried to harm the creature or because it had escaped, though neither was stated explicitly in the narrative.

Stories that involve killing, as well as tales where men simply release the captive, appear largely realistic, at least in comparison to most capture tales. Yet as the majority contain one or more fantastic elements, a supernatural quality is typical of the genre and so links these legends of ape-man capture to other sorts of "mythical" tales—including myths in the stricter sense. Of course, the idea that ape-man bodies yield magical relics is itself supernatural. All the same, we should not forget that Lio obtain equally powerful relics, including relics that cure illnesses and protect their possessors, from the bodies of certain zoologically attested animals (dugongs, for example). So, in this respect, too, there is nothing to distinguish ape-men from scientifically recognized creatures.

We're not quite finished with stories about captured ape-men. Occasionally Lio spoke of men deliberately catching ape-men. Yet, rather than obtaining relics, the captors reputedly wanted to sell the creatures. The reports turned out to be unfounded rumors—with one qualified exception, which I now relate.

In 2011 I met an elderly villager named Langga. Though Langga claimed to be a hundred years old, he seemed remarkably alert and spoke in a very animated way. Some years previously, Langga claimed, he'd captured a female hominoid (the lactating female mentioned in chapter 2) by setting a baited trap outside a highland cave. This he'd been encouraged to do by Pora, the grifter mentioned earlier. Together the two men traveled to the port town of Ende with the creature inside a sealed box, where Pora had arranged to deliver it to a purchaser. This was someone who had flown to Flores from Kupang, the provincial capital on the island of Timor, but who had likely come originally from western Indonesia. In Ende, however, the prospective purchaser started to open the box, and the creature leaped out and fled through a window, never to be seen again. As he was telling me this, Langga bemoaned the outcome, particularly as he'd never received any part of a very large payment the purchaser had promised. This and other circumstantial evidence suggest the two men did indeed take some sort of creature to Ende. But what this might have been is anything but clear. (As it happened, both Langga and Pora died in 2012, so I was never able to follow this up.)

Whatever had been in the box, the story, like other reports of deliberately captured ape-men, almost certainly reflects foreign ideas, emanating from western Indonesia, concerning "kurcaci" or "tuyul," the mischievous elves described in the previous chapter. These too are susceptible to capture. Especially on Java, people describe men purchasing elves that have been captured or otherwise obtained, to command them to steal money and thereby become rich. Not surprisingly, then, Lio who told me about recently captured "ape-men"—at least one of whom thought I was in the market for one—usually referred to them by the Indonesian name "kurcaci."

As already shown by stories about people setting fire to caves and killing hominoids trapped inside, Lio occasionally portray the same incident as part of a myth set in the indefinite past or alternatively describe it as something that happened only recently, in a more mundane setting. In some ways like stories of accidental capture is a myth told about a distant ancestor named Péro, who enjoyed supernatural benefits from an encounter with a pair of ape-men. Neither of the hominoids got caught in a snare. Rather, the ancestor detained them by way of a threat. Tellingly, though, another man listening to the story suggested that, in this case as well, the creatures had actually been caught in a trap. Told by a middle-aged man named Rani, a descendant of Péro, the story runs as follows:

> Early each morning ape-men would come and dig up tubers from
> Péro's garden. So he began to keep watch. One morning Péro saw

two hominoids enter the plot, a male and a female. When they saw him they started to flee to the mountains. But Péro threatened them with death if they tried to escape. Promising to remain, the ape-men said they would do whatever he requested. After accompanying the ancestor to his house, the hominoids asked him what work they could do for him. Péro replied that they should fell trees and clear land for a field. He gave them two bush-knives, and the ape-men went off to the forest to begin work. But Péro secretly followed them; he climbed a tall tree to watch what they would do. To his amazement the creatures put down the bush-knives, one on one side and the other on the other side of the area to be cleared, and went to sleep. The knives then began working by themselves, cutting as much forest as could twenty or thirty men and clearing an area sufficient to grow twenty large baskets of rice.

When all the wood was dry, Péro instructed the ape-men to burn it. After they finished, they asked Péro for two mattocks to turn the soil. With these they did what they'd previously done with the bush-knives; the ape-men slept while the tools worked by themselves, completing the task in just one day. The hominoids then announced that they wished to leave. Agreeing to this, Péro asked what he could give them, and they answered: a chicken, a pig, a container of maize, and a container of rice. He then asked how they would carry all these things. The ape-men said they would not carry them; Péro should place everything in front of his house and sprinkle ash in a circle around the spot. This done, the ape-men entered the circle and everything, including the hominoids, vanished.

As Rani, the narrator, later mentioned, this story is curious because ape-men do not eat rice. However, the story obviously contains far more fantastic elements than this, including tools that work by themselves, the hominoids' ability to talk—indeed, maintain a long dialogue with the human character—and their ability to disappear.

The first theme parallels captured ape-men giving cultivators like Leja the power to work gardens quickly and with ease, a benefit the creatures confer by way of their urine or body parts. In this myth it is, of course, the ape-men themselves who, by magical means, do the work. Still, Lio say ape-men also do the work when, through the possession of their relics, people are able to quickly complete agricultural work and enjoy bountiful harvests. And in the same way, the hominoid benefactors will invisibly consume the crop if beneficiaries break the taboo on taking food from granaries in which it is stored.

In this respect the story of Rani's ancestor can be understood as an instance of what anthropologists call a "charter myth," a story used to illuminate or rationalize a belief, practice, or institution—in this instance, the magical use of ape-man relics. Tools that supernaturally work by themselves also appear in other Indonesian myths. However, such stories typically don't concern figures like the Lio ape-men. For example, on Sumba, the large island immediately south of Flores, it is mythical humans who can command such tools, while in other parts of Flores people say human witches possess this ability. What's more, in the present day it is not only ape-man relics that can confer this power. People in northern Lio mentioned a stone, an object of the sort some people obtain from snakes that embody forest spirits, that when placed in the center of a field ensures that work performed on one side is magically replicated on the other side, so a cultivator's work is completed twice as quickly. Once again, then, we find that powers attributed to ape-men, including in myth and legend, are by no means exclusive to the hominoids. And taking a broader view, we've already seen how Lio believe that comparable magical powers derive from the bodies of scientifically recognized animals.

The Lesson of Fantastic Stories

What else can we learn from myths and legends featuring ape-men? To begin with the most obvious point, they certainly don't prove that the creatures are purely "mythical." As already noted, myths and legends are poor guides to what people actually believe or accept as literally true. Such stories commonly depict ordinary animals as well as human ancestors and heroes (Alexander the Great or Jesus Christ, for example) doing things—animals talking (think, for example, of Aesop's fables) or humans performing miraculous deeds—that their narrators would consider impossible in the here and now. For Lio, partial exceptions may be people regarded as witches, magicians, or sorcerers. But the point still holds. Expressed another way, myths—and, to some extent, legends too—typically depict their characters, whether animal or human, as possessing powers that in the present day belong to purely supernatural beings. And in this way myths ignore or gloss over differences between things that people ordinarily consider radically different.

That said, even the most fantastic stories can have some basis in historical (including natural historical and geological) reality.[3] Lio myths and legends incorporate practices and events that, in basic outline and shorn of possible verbal embroidery, could actually have occurred and might still occur. Examples include setting fires to caves to kill off cave-dwellers, creatures getting caught in traps, and even child abduction. In regards to this last practice it's especially interesting that, before Flores came under colonial control, the Lio used to kidnap children from neighboring districts to serve as human sacrifices.[4]

Apart from that, instances of ape-men engaging in fantastic actions seem minimal in Lio myths. A hominoid assumes the form of a particular human character in just one story (the Tana Langi cave-burning myth). The only instance where an ape-man vanishes into thin air is the ancestral myth told by Rani, and "flying" hominoids appear in none. Of course, in myths and in some legends of accidental capture, ape-men speak. Yet not only is this highly characteristic of the mythical genre, where all sorts of animals are depicted as speaking, but—depending on what Lio reports of ape-men might reflect—it is not entirely ruled out that such a creature would be capable of speech.

Just as people don't always accept things contained in their myths as literally true, so not all Lio believe that present-day ape-men can talk, change their shape, or disappear—as we saw in the last chapter. Some Lio were also skeptical about the possibility of ape-men getting caught in animal traps. Pointing to the creatures' great strength, skeptics suggested that if an ape-man got caught in a trap it could easily free itself. A few people, who credited hominoids with the ability to vanish, expressed doubts about hominoids getting trapped because, as they pointed out, the captives could simply disappear. Of course, one could raise the same objection regarding ape-men burning to death in a cave, or any method of killing the creatures for that matter, though I never heard anyone make this particular argument. All this confirms that things described in myths are frequently inconsistent with life in the present. But not everything Lio say about ape-men in the here and now is fully consistent either. One man explained the contradiction involved in vanishing or shape-shifting ape-men getting caught in simple traps by suggesting that the creatures allow themselves to be caught so they can bestow benefits on favored individuals. But this is not a widespread idea and, anyway, descriptions of ensnared hominoids weeping and pleading to be released seem to weigh against it.

While flying ape-men do not appear in any myths, it is not ruled out that the single instances of transformation and disappearing reflect the same modern media influences (especially the Javanese elves) that account for these attributes in popular descriptions of ape-men. Possibly reflecting their time-less quality, myths as stories that are constantly told and retold will readily incorporate innovations—as when narrators substitute modern firearms for far older weapons. A Western parallel is performances of Shakespeare's plays in modern costume and modern settings. Myths and legends featuring ape-men hold yet another lesson. Indirectly, they lend credibility to contemporary accounts of the hominoids, especially those given by eyewitnesses, for viewed together, different sorts of stories reveal a progression. Distinguished as tales set in an indefinite or timeless past, myths featuring Lio ape-men include the largest number of supernatural elements—like hominoids assuming the form of particular people. Stories concerning historical human figures, and thus distinguishable as legends, include fewer. And as we'll see in the following chapters, secondhand accounts of twentieth-century sightings, as well as putative eyewitness reports, include fewer still. In fact, with a few noteworthy exceptions, most eyewitness reports appear thoroughly naturalistic.

5

WHAT SOMEONE TOLD SOMEONE ELSE, OR SECONDHAND STORIES OF APE-MAN ENCOUNTERS

There's a slight but important difference between secondhand stories and the capture tales discussed in the previous chapter. The capture stories, of course, also concern people who supposedly encountered ape-men. But only in the reports I discuss in this chapter did a reputed observer tell the story directly to a particular person who, in turn, narrated it to me. In no case was I able to consult the original sources, usually because they were already dead.

Encounters in Childhood

I start with a story that has a decidedly mythical air. By all indications it reflects a dream, specifically something dreamt by the original teller. Niwa, a middle-aged resident of a highland village in the Mego region, related how, as a small boy, he and his elder brother often slept together in a field-hut. One night the brother woke up screaming that a pair of ape-men, a male and a female, were about to lift Niwa from his sleeping mat. The account thus recalls the theme of hominoids abducting children, present in two tales discussed in the chapter 4. But whether this was the intention of these hominoids Niwa couldn't say, and his brother had died before I heard his story.

Interestingly, Niwa's brother reported to the boys' father that "two monkeys" had approached the younger boy. The father, however, interpreted the incident differently; he argued that the intruders must have been ape-men, as monkeys would not be bold enough to enter a field-hut. (Whether ape-men would be so bold either is another matter—and anyway, Niwa's brother almost certainly dreamt the incident.)

During the same conversation, Niwa related another episode where ape-men had approached him while he slept, this time with an evidently more benevolent purpose. When he was about eight years old, Niwa was assigned to guard a field but eventually dozed off, in the open. As he slept, rain began to fall. But he was kept dry by an ape-man that carried him, still asleep, to a nearby field-hut; so when he later awoke he found himself inside the hut. How the sleeper knew an ape-men, or anything else, had carried him away is quite unclear—unless, of course, he saw it while asleep, which is to say, in a dream.

It's also worth noting that when Niwa told his parents about the incident, he suggested that a spirit had carried him to the hut. But they contradicted this, saying that it hadn't been a spirit but a "living being," which they further identified as a lai ho'a (ape-man).

A theme in folklore the world over, strange beings (often fairies, spirits, or souls of the dead) taking pity on and helping youngsters is not confined to Niwa's evidently imaginary encounter with ape-men. In 2014, the same year I met Niwa, the owner of the house where I was staying repeated an at least third-hand story about a man named Nonga. Nonga was an orphan who was raised by cruel grandparents. Every day they ordered him to guard their field against monkeys, close to a dense forest that, by this time, had come under government protection. Feeling sorry for the boy, one day an ape-man appeared and, after asking him to tie up his dogs (as hominoids are afraid of dogs), offered to guard the field in his place so he could go home. (How the creature conveyed this message, the narrator could not say.) Nonga was scared he would be beaten if he arrived home in the middle of the day. So the ape-man told him to take the stick his grandparents used to beat him, tear off a sliver, and hold this between his teeth; then he would feel no pain. Some time later, Nonga went to live with the hominoids in the forest for several days. Later still, he became a mystical practitioner, or "man of power."

Seké's Tale

Although a person going to live with ape-men is unique to Nonga's story, it almost goes without saying that this tale, presumably told originally by Nonga himself, serves to illuminate and support his claim to exceptional powers. At the same time, hominoids offering to guard cultivated fields recalls the mythical exploits of the ancestor Péro summarized in the last chapter, as well as accounts of how people obtained beneficial relics from ape-men. The myth of Péro is even more clearly echoed in another secondhand account, which I now relate.

Describing the experiences of a man named Seké, the story is one of four that concern ape-men visiting human dwellings, specifically field-huts near isolated gardens surrounded by forest. With one exception, all contain details of the hominoids' physical appearance that narrators claimed they'd heard directly from their sources. (The exception was a legend I include for the sake of comparison, which the storyteller had heard from his long-deceased parents.) The narrator of Seké's tale was Goda, the elder who told me the ape-man origin myth summarized in the chapter 4. Goda heard it from Seké in 1987 or 1988, by which time Seké was already very old, and the events he recounted had occurred over ten years previously. Both Seké and his wife had died by the time Goda related the story. They'd left no descendants—at least none who knew of their experience. According to Goda, shortly after the events, the couple told a number of people their story, which he described as causing quite a stir at the time. Yet, apart from Goda, no one I met some forty years later could recount the details.

Seké and his wife were living alone in a field-hut surrounded by forest, in the region now known as Kali Wajo (actually the name of the river that flows nearby). Their garden was close to the hut, but at a higher elevation and not far from a cave reputedly occupied by ape-men. One day, a pair of ape-men, a male and a female, approached the dwelling. This was about the time people had begun planting maize, possibly in October. Seeing the hominoids, the couple became frightened. They'd never seen such strange creatures before, so they ran away. When they returned, the ape-men were gone. But several months later, in January or February, just as the corn was ripening, one of the "animals" (as Goda called them) reappeared.

One evening after sunset, Seké, who was alone at the time, lit a fire in front of his hut, as people commonly do to keep warm. (Typically field-huts are too small to accommodate a fireplace.) He began calling out in the direction of his elevated maize garden, a usual way of scaring away unseen wild animals that come to feed on ripening crops. According to the direction of the plot, cultivators will cry out, for example, "hoo up there" (*huu ghéle*, where *ghéle* means "upslope" or "toward the interior"). The cultivator might then repeat the cry, mentioning other directions from which animals might come. Hearing a human voice the creatures should run away, but on this occasion something echoed Seké's call. This happened several times. After that, Seké saw a single ape-man approaching his hut, which he recognized as the male of the pair that had come previously. So he recited a conventional declaration uttered when confronted with an unfamiliar visitor: "If you come with good intentions you may approach, but if your intention is bad we will not receive you."

To this the hominoid responded: "If you receive me, my purpose will be good; I'm hungry and have come to request food." So Seké gave the ape-man food, and when it finished eating, it told him he didn't need to guard his maize field as they (assuming the female hominoid was included) would watch over it until harvest time. The man and his wife, the ape-man further explained, were no longer young and should simply rest, warning that if they continued to exert themselves they might ail and die. All the hominoid asked for in return was food, to which Seké readily agreed. He added that until the harvest, he would feed the ape-men whenever they (implicitly the male-female pair) requested—so long as they never bothered him again. And, indeed, that was the last time Seké and his wife ever saw them.

After recounting his story, Goda related what Seké had told him about the hominoids' appearance. Although looking like monkeys, the creatures were in many ways more like humans. In fact, when Seké first saw them he thought they were humans, although of a strange kind. They had no tails; their faces were wider than a monkey's; and, though described only as "rather flat," their noses differed as well. The body hair was short and sparse. The chest hair was either absent or sparse, and there was none on the undersides of the arms. Hair grew on the calves, upper legs, and buttocks and around the genitals, but nowhere was this particularly thick or plentiful. In response to my question, Goda said there was not much hair on the face and none on the forehead. The feet were small and "thin" or "flat," possibly meaning the instep or arch was low or nonexistent, or that the hominoids had flat feet. For some reason, Seké mentioned the creatures' calves, which he described as humanlike. The female of the pair (which Goda referred to as a *metu*, the Lio term for female animals) had small breasts. When I asked about height, Goda immediately responded "one meter," though he added that Seké "knew nothing of meters and centimeters" and had indicated stature only with his hand. Goda's informant never mentioned the color of the body hair nor, apparently, anything about the head hair. But he described the skin color as dark, like that of local people.

Though more detailed than most, this description corresponds to the way Lio generally portray ape-men. And despite Goda's confirming that he'd heard all this from Seké, I'm not entirely convinced that some doesn't derive from another source, or simply Goda's own knowledge. For one thing, such a naturalistic description sits oddly with seemingly fantastic details of the account, such as the creature's speaking to the cultivator, apparently in his own language, and requesting food in return for guarding Seké's ripening corn. What's more, ape-men performing agricultural tasks for an elderly cultivator obviously recalls more fantastic tales, especially the myth told by Rani about his ancestor who detained two thieving ape-men. In fact, Goda had accompanied me when I visited Rani to record the myth. Yet the details of the myth and the elderly cultivator's account are not exactly the same. More specifically, Seké's tale contains fewer fantastic elements and makes no mention, for example, of his hominoid visitors transforming, disappearing, or surrendering urine or body parts. To be certain, I asked Goda whether Seké ever received relics from ape-men, and he said he did not.

Mbira's Story and the Legend of Wanggé

In several respects the following story, concerning a man named Mbira and also recorded in 2014, is similar to Seké's. But the location is different, the time is more recent and, more importantly, this narrative seems thoroughly naturalistic. Mbira had died some three years previously. The narrator was his sister, an uneducated village woman in her late sixties named Wonga (see figure 5.1). Mbira told her about his encounter with an ape-man on the day after it occurred. So far as I could determine, this was about five years before Wonga related the tale to me.

At the time of the encounter, Mbira was lodging in his field-hut, in a mountainous location and a short distance downslope from his maize garden. One evening, just after nightfall, Mbira lit a fire in front of his hut to keep warm. Afterward he began periodically calling out "hoo" to drive away pigs and other invasive animals that might have entered his garden—in the same way Goda described Seké as doing. A little later, something began answering his calls. At one point Wonga indicated the return cry was "hoo," but she later described it as sounding like "ooo-ko-ooka doo." (This is not a sound generally attributed to ape-men, and I never heard of it again; but in any case, it doesn't correspond to any phrase in the Lio language.)

Mbira then saw a hominoid figure coming down the slope toward his hut. As the moon was bright, he was able to see it clearly. Mbira recited, "If you are bringing something good, I am willing to receive it; if something bad, then turn away" (just as Seké had done). The ape-man kept coming, making no sound, vocal or otherwise. It then stopped within a meter or two of the hut (4 to 6 feet), and when it saw Mbira—evidently it had not heard or understood his declaration—it immediately turned around and hurried away. Wonga's brother never saw the hominoid again, nor did he ever have other experiences of ape-men.

Although effectively ending here, one interesting aspect of this account is the abortive nature of the creature's visit. When it saw the man, it fled from fear. Unlike Seké, Mbira received no benefit from the creature, nor did Wonga say that it spoke. The ape-man's appearance also scared Mbira, though subsequently he suffered no illness or other negative effect. Wonga could not say why the hominoid might have approached Mbira's hut, for example, whether it was looking for food or was attracted by the fire (a possibility suggested by another story I discuss below).

As in Goda's rendition of Seké's story, the narrator gave details of the visitor's physical appearance. It was naked and generally human in form, and it stood erect and walked bipedally. Wonga described the hominoid as "small-bodied," or thin, but it seems not to have been particularly short. In fact, she thought the figure might have been no shorter than she was. As Wonga was an old woman and apparently smaller than average, this could mean as short as 1.45 meters (4 feet, 9 inches). Measurements taken in two regions of Flores in the 1920s revealed average heights for adult women of 1.48 and 1.501 meters (4 feet, 10 inches and 4 feet, 11 inches).[1] That said, I was unable to clarify whether Mbira himself had ever said anything about the creature's height.

The face of the hominoid visitor was "like an extremely ugly human's," and the neck was "quite long"—a feature I never heard in any other account. It was the creature's ugly face, Wonga said, that gave Mbira a severe fright. The eyes were not significantly different from a human's, and the skin was "dark," like local people. Answering my question, Wonga thought the body was not especially hairy, but she qualified this by suggesting that Mbira had been so drawn to the face that he'd paid less attention to other parts of the body. Even so, she said the creature's hands or fingers were (proportionally) longer than a human's, as were its nails.

Mbira had not described the length or color of the head hair, nor the mouth or teeth. Nor did he mention a tail—something he might have seen as the creature turned and moved off. More surprisingly perhaps, the man said nothing about the sex. But what's most distinctive in this account is the way the creature walked. Mbira had described it as "hopping" from one foot to another and swaying from side to side. A man who was present during my conversation with Wonga illustrated this, to the old woman's approval, in a way that suggested someone walking barefoot on a hot surface or walking on skis. The demonstrator also characterized the movement as "walking by hopping or jumping."

I'll have more to say about this peculiar walk later on. But whatever its significance, taken as a whole, the story sounds realistic. If the height estimate and relative hairlessness are accurate, there's little to distinguish the figure Mbira saw from a small and ugly naked human being. On the other hand, several details are reminiscent of an apparently older tale, more accurately classified as a legend, about a man named Wanggé. First recorded in 2015, it was narrated by Nuwa, the elderly man we met in chapters 2 and 3, who before his death in 2017 was a leading figure in his village. Nuwa was a lively, assertive, and intelligent person, not always easy to question as he would often change the topic. But he was always friendly and generous with his time.

Wanggé belonged to the generation of Nuwa's great-grandfather, and Nuwa heard the story from his parents. A bachelor his whole life, Wanggé suffered from a skin disease that left him scarred and unattractive, which might be why he had always lived alone in a hut near his garden, again in a place surrounded by thick forest. One cold evening he lit a fire on the ground in front of the hut to keep warm. Some time later, he heard a voice calling out from a higher-lying location, "Friend Wanggé, friend Wanggé, we are cold." Wanggé saw ape-men coming down a hillside, heading toward his hut. They wanted to warm themselves by the fire. There were two or three of them—either a male and a female or two males and a female. Mulling over how he could get rid of them, Wanggé went inside his hut, closed the door, and kept quiet. But the ape-men knew he was there. He heard one exclaim, "there's the smell of Friend Wanggé," indicating they had caught his scent. Then he had an idea. Wanggé had a dog, a bitch that had just given birth. As the hominoids came closer he released the dog, which started barking furiously and caused the ape-men to flee in disarray.

Nuwa insisted that these events had "actually happened" and were not just a "fairy tale" (Indonesian "dongeng"). All the same, he expressed surprise that the ape-men would have known Wanggé's name, wondering where they could have heard it. In fact, this is the only story I recorded where a speaking ape-man addresses a human by name. Retold a year later, in 2016, Nuwa's second version of the story was similar to the first but differed in a couple of particulars. On this occasion he said there had been just two ape-men and both were male. Also, whereas in the earlier version he'd implied that the dog was inside the hut with Wanggé, in the second telling the bitch and her pups were kept in an enclosure outside. Nuwa then explained that when the ape-men arrived at the hut they didn't yet know the dog was there. But all of a sudden, Wanggé opened his door, roused the dog, and set it on the unwelcome visitors. (Nuwa then mentioned, as I'd already heard many times, that ape-men are very afraid of dogs.)

Actually, the detail of the dog may be one of the less credible parts of the story. As I know well from experience, Lio dogs will bark incessantly at any unfamiliar visitor and often begin barking before a person gets anywhere near a dwelling. Also, a bitch with pups should be especially nervous with strangers. And since Lio credit ape-men with a good sense of smell, they should have smelled not only Wanggé but his dog too. But this is perhaps taking a stylized folk narrative too literally, and one can imagine a more factual basis in a dog simply chasing away unknown nocturnal visitors.

While relating the tale, Nuwa mentioned several details of the hominoids' appearance. Given the age of the story and the fact that Nuwa didn't hear it directly from Wanggé, it is likely that, rather than forming part of the original narrative (or the version Nuwa heard from his parents), these details derive from the narrator—or do so to a greater extent than, for example, in the story told about Mbira. Yet not all features mentioned were typical of Lio representations of ape-men. At different points in our conversations, Nuwa variously described the creatures as "short," "almost the same [height] as humans," and "a bit shorter" than people. This sounds rather like Wonga's account of Mbira's experience. However, Nuwa responded positively when I later mentioned that I'd heard ape-men described as standing just over a meter, which would definitely be diminutive by local standards, and he went on to compare the height of the creatures (apparently ape-men in general rather than the specimens that visited Wanggé) to the height of figures he'd seen on television. These turned out to be the western Indonesian "elves" (kurcaci or tuyul) discussed in chapter 3.

Nuwa described ape-man faces as "like animals, monkeys" or "forest people" ("orang hutan," perhaps referring to orangutans). Apparently specifying the specimens Wanggé saw (at least when he said these included one female) Nuwa said male ape-men are hairy-bodied whereas the females are less so and may be no hairier than local people. Referring more definitely to Wanggé's visitors, he also described the creatures as looking "old." Nuwa thought the head hair had been short and not much longer than the body hair. Both the body and head hair were "black" or "dark."

Although evidently much older, the story of Wanggé closely resembles not just Mbira's story but also Goda's account of Seké's experiences. All three men were visited by ape-men while occupying field-huts close to cultivated gardens surrounded by forest. In two instances the gardens were located at a higher elevation than the huts, though this arrangement is not unusual in Lio. Before their hominoid visitors arrived, both Seké and Mbira were calling out in the night to drive off animals. Wanggé was not so engaged. Yet the fact that the ape-men called to him from somewhere higher than his hut, as did the creatures that responded to Mbira's and Seké's cries, reveals another parallel with the other two stories.

All in all, the similarities suggest we may be dealing with a conventional form of narrative into which different informants have inserted different details regarding specific characters. Still, the story of Wanggé is most similar to Mbira's tale, and this is both the most recent and the most naturalistic. Wanggé's story contains the only clear reference I recorded to hominoids coming to warm themselves by people's fires. Yet this may be implicit in the stories told by Mbira and Seké, where ape-men approached field-huts in front of which these men, too, had built fires. Interestingly enough, hominoids attracted to manmade fires also appear in stories from other parts of the world—from the neighboring island of Sumba to Mongolia to nineteenth-century California.[2] Finally, in Mbira's story and the legend of Wanggé, the ape-men were physically similar. Neither was particularly short; in fact, the height mentioned by Mbira's sister exceeded all others I recorded. Also, while in Nuwa's account of Wanggé's experience only female ape-man had relatively hairless bodies, the creature described in Mbira's story (whose sex was not specified) was also not particularly hairy.

Other Hominoid Encounters

Differing in several respects from the foregoing narratives, I recorded two other secondhand accounts of ape-men seen near human habitations. The first concerns a man named Ligu who—like Seké, Mbira, and Wanggé—encountered an ape-man approaching a remote field-hut. This was told by Ligu's son, Muda, who was sixty-eight years old at the time. The reputed incident occurred in the early 1950s.

One night Ligu was guarding his maize field. In the evening he heard a sound he thought was a wild pig; then later, around midnight, a humanlike figure, an adult male with fully developed genitalia, appeared in front of Ligu's hut. It was naked, and just over a meter in height—about the same size as a small boy the narrator pointed to, who stood 1.1 meters (3 feet, 7 inches) tall. Its body was covered in sparse hair, and the face, including the chin, resembled a monkey's. Ligu asked, "who are you," and the hominoid responded with exactly the same question!

Though the details could be clearer, the hominoid apparently remained for a time by Ligu's hut. Ligu prepared to shoot it with an arrow, but fearing there might be a serious consequence, he hesitated, and the ape-man grabbed the bow from his hand. The hominoid then ran off, so Ligu summoned several companions to pursue the creature. Though they searched all night, they could not find it. Then, on the following day, the searchers discovered the ape-man had taken refuge in a nearby cave. The men gathered a large quantity of palm fiber, placed this inside the cave, and set it alight. However, the creature was able to escape through another hole. At some point someone grabbed the hominoid by the arm, but it broke free and fled; even five men could not restrain it. On another occasion—whether before or after the cave-burning was not made clear—the same ape-man gave Ligu a wooden cudgel with magical properties.

Including a speaking hominoid visiting a field-hut, a cave-burning episode, and the transfer of supernatural benefits by way of a material relic, Ligu's story reads like a potpourri of mythical and magical ideas regarding ape-men. Considered as a possible account of actual events, therefore, it seems less credible than other stories about ape-men voluntarily visiting human habitations. Nonetheless, the narrator's description of the subject's physical appearance sounds completely naturalistic, and all but one of the accounts reviewed so far—Mbira's tale—contain at least one fantastical element.

The next account is even more realistic. This is perhaps significant since it was first related to me as an eyewitness report—as it may well have been. The narrator was Reku, a fifty-year-old man from a mountain village some distance above Moni, a roadside settlement that now provides services to tourists visiting the famous three crater lakes atop nearby Mount Kelimutu. A school teacher currently resident in the north coastal town of Maumere, Reku was introduced to me in 2015 while I was visiting his elderly mother-in-law to inquire about another matter. The following summarizes what he told me on our first meeting.

One evening between six and seven o'clock, when he was about twenty and still living with his parents on the slopes of Kelimutu, Reku opened the front door of their dwelling to see an ape-man walking through a nearby maize garden. Apparently alarmed by the opening door, the creature moved away quickly. It was erect, bipedal, and over a meter tall (3 feet, 3 inches), and the head looked bald. Because it was visible only from the back, Reku couldn't say whether the face or body were hairy. But the creature did appear "strong," which Reku explained to mean thickset and not thin or slight. As it moved through the cornfield, it held its arms forward, pushing the stalks aside. At this time the maize was over a meter in height and approaching 1.5 meters (4 feet, 11 inches), and the figure "was almost as tall as the corn." Though it was getting dark and misty—a condition common year-round at these high altitudes—Reku said he could see the hominoid because the field was very close to his house.

When I interviewed him a second time, Reku said the ape-man was walking toward him when he opened the door but then immediately turned around and entered (or reentered) the garden. But if so, it's not clear why he didn't see more of the face. When I asked whether the figure might have been an ordinary human, Reku simply responded that it was very short and bald—apparently meaning that he took the baldness as a sign the hominoid was mature or elderly yet too small for an adult human. Reku never saw the creature again, nor did he hear of other people nearby seeing anything similar, though in the same conversation he mentioned that someone had recently found village pigs that had been killed.

It was at the end of this second conversation that Reku, after previously representing himself as the sole witness, suddenly asserted that it was not he but his late father who'd seen the ape-man and had described the incident immediately afterward. I didn't find this attribution entirely convincing, partly because, on this occasion, Reku had been called away from another task to speak to me, needed to get back to it, and clearly wanted to bring our conversation to a close. It's also possible that, even though at the beginning he didn't seem particularly reluctant to discuss the topic, by this point Reku wanted to distance himself from the incident. Be that as it may, I found his story quite compelling—and comparable to sighting reports detailed in later chapters.

More definitely secondhand is a report of an experience claimed by a young man, about thirty years old, named Tiwa. According to Boka, the narrator and a leading man in a mountain village in central Lio, Tiwa was too embarrassed to tell me his story himself because, unlike most people his age, he was illiterate. So Boka, a man who regularly hosted me in his village and assisted me in other investigations, had Tiwa describe the experience to him the day before he related it to me. Obviously, the arrangement was not ideal. But Boka was a man I'd always found trustworthy, and besides, he mentioned that he'd heard Tiwa's story a month or two after the incident occurred.

Some ten years prior to 2016, when Boka recounted the tale, Tiwa and his now-deceased father were returning home late one night from another village after a gambling game, when they came across a strange creature sitting on a large boulder. From a squatting position the creature "jumped away" to a higher spot, "holding its hands behind its head" as it did so. On seeing this, Tiwa's father, who was walking behind, pulled Tiwa back and stepped in front of him. Being ignorant of such things, Tiwa asked what it was, and the father identified it as a lai ho'a (ape-man). The thing was small, like a child of five or six years. (Boka also compared the height to that of a nearby child, who stood 1.05 meters [3 feet, 5 inches] tall.) Tiwa compared the face, though generally human in form, to that of a "grasshopper." Though the comparison could conceivably reflect the jumping movement, Boka thought Tiwa meant that the hominoid appeared gaunt, with sunken cheeks. The body, too, was thin. The head was bald on top, with hair growing only at the back.

The grasshopper comparison is interesting because grasshoppers also have relatively large eyes, but Tiwa never mentioned the eyes. In response to my questions, Boka provided further details of Tiwa's encounter. It occurred near a stream at a spot distant from any settlement, when the moon was bright, thus allowing a reasonably clear view of the subject. When I asked about the legs, Boka thought these too were humanlike. Tiwa had said each foot had only two toes, but his father had inferred this from footprints he later found nearby. The arms too were humanlike but "short"—though in comparison to what, the narrator could not say.

In view of the several peculiar (though quite specific) details of the subject and the fact that Tiwa never mentioned body hair, it is unlikely that Boka would have concocted this story from his own general knowledge of ape-men. He would have had no particular reason to do so. And Boka had never seen an ape-man himself.

Kowa's Confrontation and Accounts of Ape-man Killings

Like Tiwa's tale, the next story differs from most secondhand accounts because it doesn't feature ape-men visiting either dwellings or cultivations. It is also set in a far more distant past. The story concerns a man named Kowa who, perhaps around 1910, had a brush with two ape-men in a forest while traveling from his village to his garden. It was told by Kowa's fifty-nine-year-old grandson, named Dugo, who remembered his grandfather from the time Dugo was about twelve years old. Kowa died when Dugo was in his twenties, by which time, Dugo thought, Kowa was ninety years old. Dugo estimated the old man's age at the time of the incident as around twenty-five.

Early one morning Kowa was walking alone on the way to his garden, following the course of a stream, when he came to a place where there is a high waterfall. On a cliff near the top of the fall, he saw what from a distance looked like small children. They were "jumping about," apparently playing, at the edge of the cliff. Fearing they might fall off the cliff, Kowa shouted out, warning them of the danger. But they ignored him. Kowa then climbed up to the spot, but when he arrived the hominoids grabbed hold of him. Despite their small size, they were extremely strong. Kowa was thus unable to free himself, and in the ensuing struggle they threw him down the slope. When he finally recovered—apparently, he had lost consciousness for a time—the hominoids had gone. It was only then that he realized they could not have been human and must have been ape-men.

When he fell Kowa broke his lower jaw and remained disfigured for the rest of his life. His face was "crooked," so water would dribble from the corner of his mouth whenever he drank. From this encounter, Dugo suggested, his grandfather may have become a "man of power;" he was well known as a mystical practitioner, and—unusually—he even professed to be a witch. Still, the idea that the hominoids were the source of the older man's powers was only the grandson's speculation. Answering my questions, Dugo said his grandfather had described the ape-men as naked, tailless, and hairy; he added spontaneously that the body hair was sparse and grew only in certain places, including the chest. But it is not clear if the old man had actually related these details or whether the younger man was drawing on his own impressions of ape-men. And when I spoke to Dugo the following year (2017), he was not sure his grandfather had mentioned whether the hominoids were hairy-bodied or not.

As told by his grandson, Kowa's story seems quite realistic. Dugo was one of a number of Lio who spoke of ape-men as able to change shape. Yet neither this nor any other supernatural ability is mentioned in the story, only the hominoids' great physical strength. As such, the tale is one example where an older representation, in this instance dating to the early part of the twentieth century, appears more naturalistic than more recent stories of ape-men. Another may be Wanggé's story as related by Nuwa.

Kowa's story is unusual insofar as it describes ape-men physically attacking and permanently injuring a human protagonist. Another, rather briefer, secondhand report had a hominoid killing a human. In the early 1940s (during the time of the Japanese occupation) dogs belonging to an unnamed man chased an ape-man that had entered his garden into a tree. The man then struck the hominoid with a stick. But, unaffected by the blow, the creature struck back at the man, "breaking all of his bones" and ending his life. It then proceeded to eat his corpse—the only reference I ever recorded to an ape-man consuming human flesh.

By contrast, several stories describe people killing ape-men, as, of course, do myths and legends reviewed in the previous chapter—for example, by burning caves with hominoids inside. We might also recall the ancestor Péro threatening two ape-men with death and Ligu's thwarted attempt to shoot an arrow at a hominoid visitor. Rather more prosaically, a man in central Lio told me about a (now-deceased) highlander whose dogs chased and killed an ape-man, perhaps in 2001 or 2002. From the carcass, the dog owner, a mystical practitioner, retrieved one or more bones, which he retained as magical relics. Specifically, he would carry these on his person to render himself invisible.

Another secondhand report described an inadvertent killing of an ape-man. The narrator was Sipo, a young man employed as a modern travel guide who'd heard it around 2010 or 2011 from a man named Tuka, whom he'd met in a highland village in northern Lio. Sipo was not entirely sure whether the killer had been Tuka or someone else. The incident took place some years prior to their meeting, but how many years the travel guide never asked.

At the time, something was stealing tubers from villagers' gardens. Cultivators suspected wild pigs. One night, a man—either Tuka or another man—was guarding his plot, armed with a bow. In the dark he heard an unseen creature he thought was a pig and so drew his bow and released an arrow. (Sipo thought the intruder might first have been caught in a snare, but he couldn't remember.) However, when the archer inspected his kill, it looked human. Although "hairy like a monkey," it had the form of a child standing perhaps between 60 and 70 centimeters (2 feet and 2 feet, 4 inches). The face also looked human and resembled the face of an old person.

Before drawing his bow, the garden owner exclaimed, "I'm going to shoot you," whereupon the victim—in the manner often described as typical of ape-men—repeated his words verbatim. Why the archer didn't infer from this that the creature was not a pig, Sipo couldn't say. And though Sipo suspected that Tuka had been the killer, the man was reluctant to acknowledge this—because, Sipo thought, he was scared of confessing to having killed something so humanlike.

Though slim on details and generally vague, Sipo's secondhand account contains themes found in other stories, including cultivators threatening ape-men with bows and hominoids stealing tubers. As regards ape-men being mistaken for other creatures and thus killed accidentally, the story also resembles a report I heard directly from an eyewitness. In 2005 I was talking about ape-men to people in the administrative subdistrict of Kotabaru, in a settlement several kilometers landward of Tuka's village, thus closer to Lio's north coast. During the conversation one man mentioned that his nephew had once seen an ape-man. As luck would have it, not long afterward the nephew, a man in his late twenties named Laka, showed up. Laka appeared nervous, even frightened, when I introduced myself and mentioned my interest. But pressed by his uncle, he reluctantly told his tale.

About two years previously, Laka and another young man, named Mité, had been guarding a paddy field at night when they heard what they thought was a water buffalo running loose. Exercising a traditional right to kill any livestock that enter cultivated fields, Mité shot the animal with an arrow. But once the carcass was discovered, the creature turned out not to be a buffalo. Or rather, as Laka expressed it, "the buffalo (had) disappeared"— a way of speaking discussed in chapter 3 when unpacking Lio claims about disappearing and transforming ape-men. What he and his companion found instead, Laka said, terrified him. While generally hominoid, the body was small and thin, no more than 10 to 15 centimeters (4 to 6 inches) wide. The limbs were like those of a monkey as were the body hair and ears. The creature also had a tail, but this was only about 20 centimeters long (less than 8 inches), far shorter than the tails of local macaques.

Rejecting my suggestion that the creature may simply have been a monkey, Laka said the head was large in relation to the body and the face was unlike a monkey's. Unable to describe it, Laka characterized the face as "unlike anything [he] had ever seen." Two canines grew from the creature's lower jaw, and it looked "vicious." Before it died, it made several sounds, including sounds like a dog, a cat, and a child crying. Laka claimed not to know what happened to the body. After he departed, though, others present suggested that Laka and Mité had buried it so they could later retrieve the bones.

The circumstances of our conversation left something to be desired. I was never able to meet Laka again in 2005, and when I attempted to do so years later, I learned he was in jail for having stolen two of his uncle's buffalo. A local man who introduced me to the Kotabaru villagers thought Laka had been frightened of me because I was a stranger and, possibly for the same reason, was especially reluctant to reveal his involvement in killing a humanlike creature. (Recall that he claimed that not he but his companion—whom I was never able to fully identify—had dispatched the thing.) It is also conceivable that, because of what he interpreted as a humanlike appearance, Laka had not accurately described its facial features, portraying these as more animal-like. In fact, nothing in Laka's demeanor inspired confidence, nor did anything I subsequently heard about the man. So on these grounds, his story remains suspect.

All the same, it is interesting that, unlike Tuka's victim, Laka never mentioned the creature he encountered echoing human speech—a difference suggesting Sipo, the narrator of Tuka's story, possibly added this detail while retelling what he'd heard from Tuka. The point, of course, applies equally to seemingly fantastical features of other secondhand stories. It's also noteworthy that, taken as a whole, secondhand accounts include fewer implausible details than many popular accounts of hominoids given by non-eyewitnesses. For example, none mentions flying or transforming ape-men, and none describes hominoids with tails. Interestingly, the one report that did speak of tails and, if only implicitly, transformation was Laka's. But, ironically, Laka claimed to be reporting what he himself had seen!

Without questioning the original observers it's obviously impossible to say whether seemingly fantastic details of secondhand accounts—or, indeed, details of any kind—were part of the original story or are embellishments added by narrators. But whatever we make of them, secondhand stories have more in common than just being secondhand. For they reveal a number of themes that, although not exclusive to such stories, nevertheless occur in several instances.

The most common is an ape-man (or ape-men) visiting a human dwelling, typically a remote field-hut (e.g., the stories told by Seké, Mbira, Nuwa, and Ligu). A variant theme is someone coming across a hominoid in a cultivated plot (Tuka, Laka)—also the setting of the myth of Péro and legendary encounters, described in the last chapter, concerning men who obtained relics from ape-men caught in traps. In a more realistic vein most human participants are described as guarding their fields at night against intrusive animals. In fact, only the incidents reported by Tiwa and Kowa don't reveal either of these themes. But just as common to secondhand stories, all involve close encounters with ape-men and in most instances a sustained interaction between hominoids and humans (for example, an exchange of words). As we'll see in the next three chapters, eyewitness reports tend to concern more distant and more fleeting observations. And almost all seem more naturalistic and, therefore, more credible.

FIGURE 1.1, LEFT: Hunter with a Flores giant rat (*Papagomys armandvillei*; female, length 66 cm).

FIGURE 1.2 BELOW: A cast of the tiny cranium of the *Homo floresiensis* holotype.

FIGURE 2.1, LEFT: The author's grandchildren, standing from left to right: 1.3 meters (4 feet, 3 inches), 1.09 meters (3 feet, 7 inches), and 90 centimeters (2 feet, 11 inches).

FIGURE 2.2, RIGHT: A Lio man with long hair tied up in the traditional style.

FIGURE 2.3, LEFT: A visiting elder.

FIGURE 3.1, ABOVE: Dugong bones and other relics for sale in an eastern Lio market.

FIGURE 3.2, BELOW: Reto's "ape-man" bones.

FIGURE 4.1, ABOVE: Goda, a mythologist and ritual specialist.

FIGURE 5.1, LEFT: Wonga, Mbira's sister.

FIGURE 6.1, TOP: Mango, a mystical practitioner.

FIGURE 6.2, CENTER: Wula and child.

FIGURE 6.3, BOTTOM: Jata near the spot where an ape-man crossed the stream.

FIGURE 7.1, TOP: Wolo.

FIGURE 7.2, CENTER: Noko indicating the location of the garden where he saw the ape-man (at the foot of the further slope).

FIGURE 7.3, BOTTOM: Dhiki with a child whose size she compared to an ape-man's.

FIGURE 8.1, ABOVE: The Ebu Lobo volcano in Nagé (the young men began their journey to the left of the scar, traveling clockwise).

FIGURE 9.1, BELOW: A Malaysian negrito holding the carcass of a white-handed gibbon (courtesy of Kirk Endicott).

FIGURE 9.2, TOP: Tipa, at his stall in
a weekly market.

FIGURE 10.1, BOTTOM: An artist's
impression of *Homo floresiensis*
based on available morphological
information (Inge van Noortwijk,
November 2021).

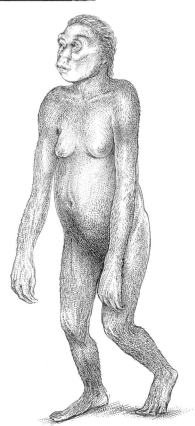

PART III

EYEWITNESS ACCOUNTS

6

FROM QUESTIONABLE STORIES
TO COMPELLING REPORTS

should perhaps mention again that I don't consider everything I heard about ape-men to be accurate or factual, and this includes eyewitness accounts. But how do we separate the wheat from the chaff—or sighting reports that contain more wheat or more chaff? Simply listing the accounts chronologically is neither useful nor entirely practical. For one thing, even the year of an encounter could not be determined in some instances. For another, there's no room to cover all sightings, even though some eyewitness reports have been described previously (for example, Laka's account, summarized a few pages earlier, and the story of the white civet related in chapter 3).

In these next two chapters, then, I present illustrative accounts by self-proclaimed eyewitnesses in what might be called an order of credibility, starting with what appear to be the most fanciful and proceeding to the most plausible. Admittedly, the arrangement is subjective, though I don't think unduly biased. Anyway, once all the reports have been discussed, we'll be able to see how well it fits several objective criteria.

Misidentified Monkeys

We begin with putative sightings that are the least likely to have involved unidentified hominoids. In at least seven of these, the subjects were monkeys or, in one instance, a civet. Probably the most clear-cut case comes from two young men who observed the subject simultaneously. As with all sightings where there were two or more witnesses, I questioned the men separately. I also spoke to one of them, Koli (born in 1979), in 2005 and again in 2015, and what he told me on the two occasions hardly differed.[1] Not only that, the details matched the report of the other man, Watu (born 1987), whom I was first able to speak to in 2015.

During the dry season of 2003 or 2004, Koli and Watu were hunting in the highlands. One day, about noon, a dog they'd brought with them barked and ran toward a large tree. Following behind, they saw a creature climbing the tree. It sat down on a branch, according to Watu, in the manner of a monkey. In fact, both eyewitnesses said the creature looked like a monkey, the main difference being its tail. This was shorter than the tails of local macaques, and Koli estimated the length as between 15 and 20 centimeters (6 to 8 inches). Watu added that the body was larger than a monkey's, about the size of a dog's, and about half a meter (20 inches) long, in Koli's earliest estimate. Koli also thought the arms looked a bit shorter than a monkey's. However, it ran from the dog on four limbs and, at one point, Watu seemed to conclude that it had been a monkey.

Why the young men thought otherwise lay in how the creature behaved. Several times they fired on it with their air rifles. Yet not only were they unable to hit it (a detail that may recall the ape-men's invulnerability to metal weapons) but, immediately afterward, the creature suddenly "disappeared." This scared the two men, both of whom mentioned the creature's apparent vanishing act repeatedly. Yet the disappearance finds a ready explanation in a detail provided by Watu, who described their target as "disappearing" just as he and his companion were reloading their air rifles. Watu also confirmed that they reloaded simultaneously, and that it was when they looked up again that the creature was no longer in the tree.

Despite these circumstances, both men maintained a supernaturalistic view of the incident. At the time they observed the disappearing primate, neither knew about lai ho'a (ape-men), and it was only when they shared their experience with family and friends that older people determined that they'd encountered an ape-man. One significance of this case, then, is the light it throws on the idea, discussed earlier, that ape-men are able to "disappear." For if monkeys (especially large-bodied, short-tailed specimens) can appear to vanish into thin air, there's no reason the same should not apply to larger primates.

Suspect Stories (Mango, Gali, and Pora)

The next several reports are more difficult to accept as descriptions of monkeys or any other familiar animal. At least it's not clear what (if anything) the witnesses may have seen. The circumstances of the sightings also raise questions, for whereas Koli and Watu described something they observed in broad daylight, all these encounters took place at night. Also, while the young hunters struck me as a pair of innocents, the character of these other witnesses is a concern, not least because all were reputed "men of power," or known mystics.

One such person was Mango (see figure 6.1), a youthful-looking sixty-year-old who not only claimed to have seen an ape-man but also boasted several encounters with the spiritual beings called "vine mothers." (In all instances he unaccountably survived the meeting, and one was atypical in that he could see the supernatural woman but she couldn't see him.) Mango's sole sighting of an ape-man occurred on a moonlit night around midnight in August or September of 2016, the year before I met him. With a companion he was descending a steep path, returning to his home in a coastal village after visiting relatives in the mountains. At a quiet spot Mango had stopped to urinate at the edge of the path when, in the adjoining forest, he suddenly saw a humanlike form streak past him, very close by, among the trees. It was heading downslope, and soon disappeared from sight.

Mango described the figure as bipedal, dark-skinned, thin-bodied, and about a meter tall. The head hair was short, but whether the creature was hairy-bodied he couldn't tell. Responding to my question, though, he was quite sure the thing was male. So far the description sounds reasonably naturalistic, but here the naturalism seems to end. For the hominoid wore a "small white singlet." Not only that, Mango mentioned that, a moment before the figure flashed by him, there had been a "sign" that he might see an ape-man. Just as he was about to urinate, a large bird resembling an eagle swooped down on his companion, who was sitting resting at the time, and it was only when Mango looked up that he saw the hominoid rush by in the forest. According to Mango, his companion also saw the ape-man. But I was unable to question the man, a schoolteacher, because some time after the incident he was posted to another part of Flores, where he died within a few months.

Further giving pause as to Mango's credibility was his claim that he tried to reach out and grab hold of the thing. This, he said, was because he'd heard that by capturing an ape-man one could obtain benefits, an idea reminiscent of the imported thieving elves (kurcaci) discussed in chapter 3. Similarly, the prior appearance of the large bird recalls the eagle that flew out of a nearby tree after nocturnal fishermen observed an "ape-man" in the form of a white civet approaching them along a forest vine. We should recall, however, that the fishermen interpreted the civet as transforming into the large bird. By contrast, Mango put no such interpretation on his experience, and if his report is given more credit, it's possible that the swooping bird (most likely an owl) was either a coincidence or a piece of embroidery. By the same token it's equally possible that the "white singlet" was a pale area on the creature's body—a feature mentioned in another eyewitness account discussed below. What's more, it also recalls what was apparently a white garment worn by a similarly small hominoid once seen, also in the Lio region and after dark, by a foreign researcher whose identity I have been asked not to disclose.

At the same time, Mango's story recalls an even less credible story related by another mystic. In 2003 I recorded a secondhand account of an encounter reported by Gali, an older man we've met before, who'd stumbled across a small group of ape-men. This occurred some twenty-five years previously, late one night on the beach near his home village—as it happens, the same coastal village where Mango lived. Sitting on the sand weeping, the hominoids were about to bury a dead female companion in the sand. As he approached them they threw sand at Gali, but he was nevertheless able to snatch the body.

When I was able to interview Gali in 2005, however, he gave quite a different account. The corpse, he claimed, took the form of a creature—some 30 centimeters (1 foot) long, with light-grayish or beigy skin (indicated with examples)—that lacked arms and had only three toes on each foot. "Hair" (*bua*, also "feathers"—the word in Lio is the same) hung down from where the arms should have been, and the body was covered in very fine "hair." What Gali saw, then, sounds rather like a bird. Other details of Gali's story also suggested birds. He described the burial party as covering the corpse with their "wings," and when he approached them, they "fled" (or "flew away") emitting cries like a large bird called *gaka*, possibly a light-colored raptor. The old man's description suggested the white-bellied sea eagle (*Haliaeetus leucogaster*), a coastal bird, light-grayish with a white head and breast, which can occur in family groups of up to three—except for the fact that I've found no evidence of sea eagles being active nocturnally.[2]

Whatever Gali may have seen, his interpretation of the creatures as ape-men was probably not unmotivated. Before his death around 2010, Gali was another mystical practitioner and, moreover, a dispenser of relics, who provided clients with fragments of what he claimed were ape-man skeletons. Some of these he said he'd retrieved from the dead creature he found on the beach. And it was one such piece of bone which, as mentioned in chapter 3, turned out to be a cranial fragment from a long-tailed macaque.

Also dealing in ape-man relics—including a bottle of urine, which he tried to sell to me—was another man who claimed to have encountered a group of hominoids. A resident of the eastern Lio village of Nua Lolo, Pora, a man we've met before, seemed knowledgeable and open when we first met in 2005. Yet because of his pecuniary interest in the topic of ape-men, I had reservations about him from the start. These were confirmed in later years when it became clear that Pora was well known as a swindler and opportunist; as shown in the previous chapter, he was prominently involved in the incident in which what was reputedly an ape-man was captured and sold to prospective purchasers from outside Flores. Aberrant details from a general account Pora first gave of ape-men included his idiosyncratic claim that his family maintained a hereditary mystical relationship with the hominoids; his possession of artifacts he said had belonged to and depicted images of ape-men (all, as I later found out, were traditional items depicting human ancestors); and his further claim to be able to conduct a rite that would cause ape-men to magically appear.

Pora's sole sighting of ape-men occurred late one night when he was returning home by motorbike on the Trans-Flores Highway in or around the year 2000. Arriving at a road bridge near the central Lio settlement of Watu Néso, he stopped his bike because he thought he heard "people on the bridge, talking." But what Pora saw in his headlamp were a number of ape-men—perhaps ten or more—in the middle of the bridge, some standing, others sitting, and some "leaping about." When they saw him some fled at great speed. But others remained, and these he "spoke to," explaining that he was "a relative of theirs" and requesting that they allow him to continue his journey safely.

Actually, Pora's description of the creatures was both detailed and mostly naturalistic. It agreed with the common representation of ape-men as hairy-bodied erect bipeds, ranging in height from about 30 centimeters (1 foot) to a meter (3 feet 3 inches)—two extremes represented within this particular group. Pora went on to describe the top half of their heads as humanlike, the eyes as "truly human," and their noses as humanlike although "flat" (Indonesian "pesek"). Their eyebrows were like humans' but thicker, and the males had "beards" on the sides of the face and below the chin. Pora also mentioned that he never saw any females among the group, or any that appeared distinctly female—a point he later reiterated.

While Pora talked to them, the hominoids remained silent. Indeed, he thought that ape-men in general, though able to communicate among themselves, did not possess a language intelligible to humans. Contrary to most accounts, however, Pora said the head hair of the creatures he observed was long and free-flowing. The claim that he saw a group of as many as ten ape-men is similarly unprecedented among sighting reports, and it also doesn't fit well with popular descriptions. Pora's statement that the hominoids were standing or sitting on a road bridge might also give some pause. A spot just beyond another road bridge was the location of a more credible nocturnal sighting by Malo, a man resident in the same village whose house was located almost directly opposite Pora's. And since Malo related his experience to relatives and neighbors directly after arriving home, it's quite possible that his report is the source of Pora's story, as we'll see later on.

Claimed sightings of ape-men on roads are not unusual. In 2005 I met Woda, the elderly man mentioned previously who spoke of two kinds of ape-men distinguished by size and hair color. Woda said he'd encountered ape-men on the Trans-Flores Highway twice in his younger days, including once on the road between his home village and Watu Néso (where Pora's sighting also occurred). Both sightings took place at night, while Woda was riding in a motor vehicle. On one occasion, he claimed, two ape-men clung onto the back of the vehicle, then jumped off and ran in front, so they were visible in the headlights. It goes without saying that this account sounds more fantastical than Pora's. In fact, Woda, yet another mystical practitioner, described ape-men in a more supernaturalistic way than did the majority of Lio I spoke to.

Dreamlike Experiences (Ura, Bajo, and Rina)

Other accounts require a different interpretation from any reviewed so far, for they strongly suggest dreams. The most straightforward instance was related by a mountain villager named Ura. In his fifties when I spoke to him in 2016, Ura described how, as a lad of seventeen, he was once collecting fallen candle-nuts in a garden when three male hominoids snuck up on him and began tickling him about the waist. When he told them to stop, they simply echoed his words and continued their tickling. Several times the creatures suddenly vanished, only to reappear in a different spot. They then pushed him down a slope into a nearby stream. But Ura was somehow able to grab a length of *wulu* bamboo (an ape-man bane) and strike out at the creatures. They then disappeared for good.

While all this was going on Ura called out to his father, who was in the same garden and only about 30 meters (100 feet) away. But he never responded, and when Ura later returned home, he was surprised to find his father there rather than in the garden. Ura's encounter also included the creatures' instructing him to open his mouth and then urinating into it. Though tickling and suddenly disappearing are sometimes attributed to ape-men, urinating in this way is peculiar to Ura's story. And while urine is a valued ape-man relic, delivering it by this means is decidedly unique! But then neither do these "ape-men" sound much like the hominoids as usually described. Fully human in form and standing about 75 centimeters (2 feet, 6 inches) tall, Ura's assailants had no body hair, not even eyebrows, and their short head hair stood straight up as though cropped. In addition, their skin was lighter than local people's; their teeth were very short, "as though they'd been filed almost down to the gums;" and while otherwise naked, they wore short pants.

The most obvious dreamlike elements of Ura's story include the hominoids' sudden changes of place and his father's not responding to his pleas. Because he described his experience as something that actually happened, his account has all the hallmarks of a dream remembered as a waking reality—or what psychologists call "dream-reality confusion."[3] Lio distinguish what they see or experience in dreams from things that occur while awake, as presumably do most people. But like other non-Westerners, Lio attach more significance to dreams than Westerners typically do. Thus one name they give to mystical practitioners who engage in clairvoyance or prophecy—for example, when searching for causes of illness or affliction—is *ata nipi*, or "dreamers."

Even so, Westerners, too, are not always able to correctly distinguish whether something happened in a dream or whether it occurred while awake. Of course, the likelihood of a dream being misremembered as a waking experience depends on whether entities or incidents in dreams are deemed possible in "real life"—like running into a long-lost friend as opposed to levitating or encountering a unicorn. So the Lio conviction that lai ho'a (ape-men) are natural creatures, even though their appearance in dreams can depart from the way there are usually described, can obviously make a difference.

Another story of a hominoid encounter apparently reflecting a dream was told by Bajo, a fisherman living in the same coastal village as Mango and Gali. About four years before I heard his tale, Bajo had been awakened one night by what sounded like one or more children outside. But when he got up to look, what he saw, passing in front of a neighbor's house, was a strange creature that he identified as a "lai ho'a." An electric light was burning at the time. However, Bajo's description not only showed little correspondence with the usual image of ape-men; it didn't really amount to a coherent picture of anything at all. It was, he said, "like a kangaroo," an animal Flores Islanders now know from modern media. Or at least it moved like a kangaroo, with a sort of jumping motion. Sometimes it went on four limbs, otherwise on two, and its height or size Bajo compared to a dog's. The face was not humanlike but resembled an owl, with large eyes. The body hair was short like a cat's and multicolored, though mostly dark. It also had long nails and a very short tail. So we shouldn't wonder at Bajo's further remark that the spectacle made his hair stand on end.

Apart from the creature's bizarre appearance and the unlikelihood of ape-men appearing inside a coastal village, another reason for thinking Bajo's experience was a dream—or perhaps, more specifically, an instance of "false awakening" (where someone thinks they have woken up but, in fact, are still dreaming)—is circumstantial. The neighbor whose house the thing had passed by was Mbaku, a man we've met before. Mbaku was away at the time; his house was empty because he and his family were spending the night in a distant hospital where Mbaku's wife lay seriously ill. Immediately after Mbaku's return, therefore, Bajo told his neighbor about what he had seen, apparently because he interpreted it as an ill omen for his wife's recovery. As it happened, the woman died shortly afterward.

Actually, it was Mbaku who first told me about Bajo's story. Even years later, Bajo was reluctant to speak to me about the incident. Mbaku's description of the creature was much like Bajo's. Mbaku added, however, that when the thing walked slowly on two legs, it appeared slightly stooped, with the elbows drawn toward the body and the forearms held at an angle. Bajo had estimated the height as 60–70 centimeters (2 feet to 2 feet, 4 inches). Mbaku also vaguely mentioned that there was "something reddish" around the creature's neck, and he thought it had a face like a dog's, only rounder and flatter, or less prognathous. Bajo hadn't said anything to Mbaku about a tail. Were it not for the short or absent tail and the jumping locomotion, both Bajo's description and his neighbor's secondhand account sound rather like a palm civet, the dog-faced and long-tailed nocturnal animal described earlier. Bajo was not the only person who compared ape-men to kangaroos, especially in the way they move, but this description is anything but common.

Also set on the coast, yet another story points partly to dreams. A middle-aged fisherman named Rina related how, some ten years before telling me his tale, ape-men had several times entered his house at night and, on at least one occasion, he had seen one. This is the sole account by someone claiming to see a lai ho'a inside a dwelling in hominoid form. The visits began after he found an unusual stone, which he kept in his house and which he therefore suspected was an "ape-man stone." During one visit a hominoid had pinched his sleeping wife's cheek, leaving a mark. Subsequently, the nocturnal visitors stole cooked rice from the house, and pigs began disappearing from a nearby pen. So, angered by these incursions, Rina got rid of the stone by tossing it into the sea. The visits then came to an end.

Like the tickling in Ura's story, the cheek-pinching suggests a dream. Recall, too, that it is also in dreams that Lio typically learn about the special powers and associations of peculiar stones they find (though these typically relate not to ape-men but to encounters with snakes and other creatures associated with forest spirits). But Rina's story has an additional explanation. Several months before I met him, Rina had suffered a stroke that left him physically disabled and verbally and cognitively impaired. Besides, Rina was introduced to me by

a local mystical practitioner who was treating him at the time, and to whom
Rina had previously related his experiences with the home-invading ape-men
as part of his therapy.

Other Sightings in the Dark

Pora's earlier-discussed encounter with multiple hominoids was not the only
sighting reported by a motorcyclist traveling at night. In fact, other such
reports seem more credible, their main shortcoming being their brevity. Luka,
a man in his thirties, once casually mentioned to me how some years previously he was riding after dark around 7:00 or 8:00 P.M. on a highland
road (a road secondary to the Trans-Flores Highway) when an ape-man
crossed bipedally in front of him. Some distance past the spot, he stopped
and turned his headlamp on the creature, which remained standing by the
roadside. It was about 60 centimeters (2 feet) tall, hairy and tailless, and had
a "terrifying face." Luka's account surprised me because I'd met him the previous year and we'd discussed ape-men several times, but never before had
he revealed this personal experience.

Related by Repu, a man about fifty, another report is a little more detailed.
Repu is the person I cited previously in connection with Lio ideas about "transformation." Once, Repu had been traveling by motorbike on a mountain road,
when around 4:00 a.m, he saw an ape-man on the other side of a small roadside
stream, sitting on its buttocks with its knees drawn up toward its chest like a
human. Its sitting height appeared to be between 40 and 50 centimeters (16
and 20 inches). The creature had a humanlike face, though it also resembled
a dog's, and its body hair was dark and fine. But how Repu was able to make
out these features so early in the morning he couldn't clarify. It was on this
same journey that Repu saw the large rock that he first took to be a cow. So
perhaps he made a similar mistake with the "ape-man."

Another sighting that took place in relative darkness occurred during the
daytime but inside a large highland cave. The witness was Dapo, a man in his

midforties and a mystical healer of some repute whom I met regularly over a period of four years. Around 2001 or 2002, some thirteen years before he told me his tale, Dapo visited the cave with two companions to collect the nests of edible-nest swiftlets (*Collocalia fuciphaga*), the sort Chinese use to make "bird's nest soup." Unfortunately, I was never able to question the other two men. The oldest man, Tamu, was working in Borneo throughout my trips to Lio.

When they entered the cave, near the cave mouth the three men came across blood and a placenta, which Tamu declared belonged to a female ape-man that had recently given birth. They then went further inside, where on a ledge around 15 feet above him, Tamu spotted what looked like the back of a hominoid that was facing toward the rock wall. This he took to be the new mother holding an unseen infant to her chest. Employing a strong flashlight (the brightness of which I was later able to test) Dapo, too, saw the figure, which he described as covered in dark hair and apparently tailless. Maintaining a sort of squatting posture, it measured about 55 centimeters (1 foot, 10 inches) from the head to the base of the spine. (Fifty-five centimeters was the height of a stone wall near where Dapo and I were conversing, which he took as his measure.)

Shaken by what they'd seen, the three men did not tarry long, moving quickly to another part of the cave to search for birds' nests. They saw no bats in the cave, and anyway, as Dapo himself pointed out, what he described was nothing like a bat. Nor is a monkey a likely candidate. Not only was there no visible tail; the far smaller macaques never enter or inhabit caves.

Influenced by Tamu's interpretation of the afterbirth and the possibility of the cave harboring a female ape-man that had recently given birth, it is of course conceivable that what the bird nesters saw was nothing more than a rock formation or a shadow suggesting the shape of a hominoid. When I briefly questioned them about the matter, both Tamu's son and wife (who had not accompanied him to Borneo) said Tamu had only ever mentioned the blood and placenta and never the figure on the ledge. Dapo, by contrast, was sure he'd seen it. But then, Dapo was a supernaturalist and generally given to fantastic ideas. I don't doubt he entered the cave, but what he actually saw and how far his account reflects Tamu's interpretation is a different matter.

Dapo's tale is not the only report linking ape-men with caves. In 2005, I heard the story of a man from Sikka (the ethnolinguistic region immediately east of Lio) who was living in southern Lio around 1999. As I never got to speak to him myself, this is another secondhand story. According to the narrator, the man was about to enter a highland cave to search for swiftlets' nests—like Dapo and his companions—when he saw several ape-men emerge from the cave, perhaps as many as seven or eight. This was the first occasion the man had encountered such creatures, and he learned what they were only after he told others of his experience.

Rawi, a Doubtful Eyewitness

Of all accounts from people who claimed to have seen ape-men, one I've found especially difficult to assess was provided by Rawi, a man in his seventies I've mentioned several times before. Another mystical practitioner, now resident on the south coast, Rawi claimed to have come across the hominoids several times in his younger days. But because of his animated and rapid speech and his habit of ignoring my questions, I often had trouble linking the details to any specific encounter.

Some of Rawi's claims were, moreover, quite fantastical. Describing the creatures as extraordinarily strong, he related how he'd once killed an ape-man with his bare hands. This happened on a sandy beach, where he buried the carcass—as he said one must do to prevent it being disinterred by other hominoids. Months later, he dug up the body to retrieve the skull, which he took with him when he later traveled to Indonesian New Guinea to work as a laborer. Evidently, then, this is an example of travelers carrying ape-man relics as protective amulets. Yet when I asked him about this, Rawi claimed, quite disingenuously, that he'd never heard of such a practice. Also, some details of this story recall the tale told by Gali, another coastal villager who, before his death, had lived not far to the east.

That said, Rawi's description of ape-man appearance and behavior, though atypical in several respects, was one of the most detailed I recorded. Not only that, he gave a thoroughly naturalistic account of hominoids he said he'd observed more than once when he was in his twenties, thus about fifty years previously, near a garden he cultivated some 8 kilometers (5 miles) inland. This was on the slopes of a peak named Wolo Mota—a place I was never able to visit because the rough terrain was beyond my sixty-nine-year-old physical capabilities. At a spot near Rawi's garden was a cave with an opening just 40 centimeters (16 inches) wide inhabited by hominoids, and below this was a stream where the troglodytes would hunt for frogs. (Rawi was the only person who described ape-men as consuming frogs, and also as catching junglefowl.) Once, he claimed, his dog had chased an ape-man into a cave, but whether this was the same cave or another I was unable to clarify.

As I learned from another Lio man who'd spoken to Rawi, the old man had less credibly described ape-men as being able to enter cliff faces where there was no hole and as entering cavities that would afterward vanish or close up—apparently, like Aladdin's cave. Rawi was also one of the few people I met who claimed to be able to catch ape-men, adding they must always be eventually released and never taken away from Flores Island.

Yet Rawi's account of ape-man morphology seemed considerably more realistic. He described the creatures as mostly human in form, bipedal, and about 75 centimeters (2 feet, 6 inches) tall—much taller, he added, than monkeys. Their heads, too, are humanlike, about 10 centimeters (4 inches) in width, and include bony protuberances (Indonesian *bonggol*, "lump, bump, hump") below and on either side of the eyes. Apparently a reference to some sort of torus (or perhaps simply very prominent cheekbones), these "bumps" distinguish the hominoids from both humans and monkeys and are so thick they obscure the ears, which are quite humanlike but lie flat against the head. The forehead is "small." The body hair is not too dense and about 2 centimeters (0.8 inches) long. The head hair grows a bit longer and sweeps backward—rather than falling forward and covering the forehead, as in monkeys—thus exposing the skin between the brows and the hairline. The nose, too, is more humanlike, though rather flat.

Also in more detail than in other accounts, Rawi described the teeth as humanlike—"well-spaced" and "without gaps." More questionably, perhaps, he stated that only male ape-men have pronounced canines, more specifically lower canines, a detail recalling Laka's story (in chapter 5). Ape-men's eyes are similarly human but are surmounted by hairy brows and perhaps heavy ridges—apparently corresponding in part to the bumps below and on either side of the eyes. The eyelashes are "not too long" and, rather than curving, stick straight out. Rawi described the chin as "curved," but what he meant by this he didn't explain.

Rawi described ape-man feet and toes as longer than those of monkeys and more like human feet. The hands, too, are humanlike. Whether Rawi referred to their feet as being absolutely or proportionally long I was unable to make out. Yet proportionally long feet could fit with Rawi's further claim, echoed in other reports, that the creatures are great jumpers. When attacking opponents, he added, ape-men can spring over distances of up to 5 meters (over 16 feet) "as though they were gliding [through the air]." They then stick their sharp nails into their victims.

More questionable details of Rawi's account included ape-men's possession of short tails. These he claimed are longer in the females than in males (growing to 10 to 15 centimeters, or 4 to 6 inches). But the main difference between the sexes is the females' breasts, which are much like the breasts of human females. Male ape-men are also bolder than the females, and only the males are brave enough to descend from their highland retreats. Rather more curiously, Rawi asserted that what he described as male "leaders" of hominoid groups are distinguished by a hairy bulge about the waist. (This last feature could suggest a distended belly, which as we'll see later, other eyewitnesses attributed to their subjects.)

Despite the peculiarities, Rawi's description of ape-men's physical appearance contains so many detailed particulars that it suggests a basis in a natural creature. But could these have had a source other than his own (reputedly quite frequent) observations? That he'd heard all or part of it from someone else is possible but unlikely. An even less probable source is media depictions of archaic hominins (like *Homo floresiensis*, whose remains were discovered in 2003). Rawi was a noticeably poor man, with little education and with infrequent if any access to newspapers or television. Also, like most Lio, he was neither familiar with nor had access to the internet. Despite his reputation as a mystical practitioner, Rawi struck me as—on the whole—sincere, forthright, and honest (unlike some others of this status). And his long account of ape-men included no mention of such notions as hominoids "flying" or, though he seemed familiar with the idea, their power to transform into other creatures.

More Compelling Reports
(Malo; Bata and Wula; Jata and Ari; and Lalu and Ndori)

As eyewitness reports, the remaining stories of ape-men are more valuable than Rawi's chiefly because they all refer specifically to single sightings and all lack the more questionable details mentioned by Rawi (such as the hominoids being tailed, and females' tails being longer than the males'). Not only that, all accounts but one describe observations made by two or more people at the same time.

The exception is a story related by Malo, a resident of the eastern Lio village of Nua Lolo and once the neighbor of the previously mentioned Pora, who died in 2012. Malo is a modern man in his late forties, of high standing, and educated beyond secondary school. At various times he has worked as a government official in the north coastal town of Maumere as well as for a political party. After coming to know him over a period of three years, I consider him trustworthy (in contrast to his former neighbor!) and his story as genuine.

I first heard Malo's story in 2015 from his younger brother, who provided a briefer but otherwise similar version of the tale. A day later Malo himself, though initially reticent, agreed to tell me about his experience—all the while stressing that he couldn't be entirely certain about what he had seen. One night in 2003 or 2004, around 1:00 A.M., Malo was returning home by motorbike from Maumere. He had just passed a road bridge when, at a lonely spot with forest on either side, he saw something in the middle of the road. So he stopped. About 20 meters (65 feet) away, illuminated by his headlamp, two hominoids sat facing one another, making murmuring or mumbling sounds as though "quarreling." Although his lamp shone directly on the creatures, they did not run away, thus Malo became very afraid. Strengthened by his Christian faith, however, he crossed himself, started up his engine, and resolved to run over the hominoids if necessary. As he advanced, the creatures got up and ran away on two legs, one to the left side of the road and the other to the right, both disappearing into the forest and still vocalizing as they ran.

Malo estimated the hominoids' height at 40 to 50 centimeters (1 foot, 4 inches to 1 foot, 8 inches), but he couldn't be sure if this was how they appeared while sitting or standing. Even so, he was certain they moved bipedally—I checked this several times—and that their bodies were covered in brownish hair, darker than local monkeys. Though he saw the creatures only in profile—and so was unable to describe their faces—he never noticed any tails. Introducing his story, Malo first said the creatures were either monkeys or "those things," meaning lai ho'a (ape-men). But he later concluded that the creatures could not have been monkeys, or any other quadrupedal animal, because they ran away on two legs and because their voices were different from monkey vocalizations. Though he didn't bring up either point, he might have added that Florenese monkeys are not active at night and that they have very long and noticeable tails.

Since then, and despite regularly traveling after dark, Malo has never again seen creatures of this sort. Throughout our conversations, he appeared genuinely puzzled by his experience. When he got home Malo immediately described what he'd seen to his family, who were holding a celebration with neighbors in attendance. As mentioned, one of these would very likely have been Pora, who may then have used Malo's story as a basis for constructing his own.

Like Malo's experience, the sighting I now describe took place at night. But it was reported by a married couple who claimed to have seen several hominoids simultaneously.

In 2016 I was visiting a highland village to speak to a man I'd met eleven years before. While there I came across Bata, a man who mentioned that his wife had seen several lai ho'a (ape-men) some years previously at the back of their house. Unfortunately, the wife, named Wula, was away at the time, but Bata, a rather terse though not unfriendly individual, agreed to tell me what he remembered of her story. Most likely Bata thought that was the last he'd see of me. However, the next year, 2017, I called on Bata again, and as luck would have it, Wula was also at home.

It's important to note that when I arrived at their house neither husband nor wife had any idea I was coming and so could not possibly have rehearsed or coordinated their accounts. Wula turned out to be a forthright and lucid woman of thirty-seven or thirty-eight, just a few years younger than her husband (see figure 6.2). Among the first things she told me was that Bata, too, had observed the hominoids—a fact the husband readily acknowledged—as had his late father. And, though more detailed, Wula's story largely corresponded to what her husband had told me the year before.

Wula had been the first to spot the creatures, a detail that could explain why Bata had previously attributed the experience specifically to his wife. Between ten and fifteen years previously, one morning about 4:00 A.M., when the moon was low over the horizon, Wula arose to relieve herself at the back of the house. After opening the door, however, she was stopped in her tracks by the sight of two small hominoids near a ra'u tree (Dracontomelon edule) some ten meters (33 feet) away. The figures walked around the tree, stopping here and there, picking up fruits and eating them. Known in English as the Pacific walnut, the fruit of this tree is commonly eaten by children throughout Southeast Asia. Also, the name is misleading, for unlike a true walnut and like a plum or cherry, the fruit does not have a hard shell and requires neither shelling nor peeling.

Evidently the sound of the opening door did not disturb the fruit collectors, who seemed unaware of the woman's presence. Wula quietly went back into the house and roused Bata and his father, explaining that there were what looked like "small children" behind the house. Apart from their small size, Wula probably called the creatures "children" because, like real children, they were snacking on fallen ra'u fruits. But her father-in-law asserted—reasonably enough—that small children could not possibly be abroad at this hour, so they must be lai ho'a (ape-men)—a name Wula had not heard before.

All three residents went to a back window to peer out at the diminutive hominoids. After watching them for a while, Bata's father roused a sleeping dog to drive the little ape-men away—because, in Wula's words, "such things are not good." The creatures then fled in great haste, on two legs, down a bank behind the house and disappeared from view. Wula closed her account by remarking how, in spite of their small size, the creatures appeared able to move very quickly.

Actually, in the previous year Bata, Wula's husband, had said there were no dogs in the house at the time. But Wula's account seemed more likely—and Bata did not demur. Both husband and wife described the figures as generally humanlike, standing erect, and walking on two legs. Though the creatures were naked, neither Wula nor Bata noticed any tails. Nor, because it was quite dark, could Wula comment on either their hair or their gender. A year before, Bata had said the creatures were hairy-bodied, whereas on this occasion he seemed to agree with his wife. Yet in 2017 he also stated that he'd seen head hair, "like the hair of a newborn infant . . . black but not too long."

Both spouses described the skin as the same color as that of their own young children. Referring to their four-year-old son, they estimated the creatures' height at about 60 centimeters (2 feet) or coming up to the shoulders of the boy, who stood about 80 centimeters (2 feet, 8 inches). Bata then remarked that, unlike small children, their heads were not large but in proportion to their bodies, and he agreed that they could have been adults. Though the informants' responses were vague, the creatures apparently were quite slender or, at least, did not appear "stocky" or "muscular."

In 2016 Bata had referred to three or four figures, whereas Wula said there were only two. However, this discrepancy, like one or two others, is easily explained by the poor lighting. At the time there was no electricity in the village, so the three observers may have been able to see little more than silhouettes against the light of the setting moon. Yet both spouses were sure that the figures stood erect and moved on two legs, and as both pointed out, for this reason they definitely could not have been monkeys. As Lio consume the fruit of the Pacific walnut, it's not impossible that the tiny collectors were (modern) humans. But as both Wula and Bata pointed out, agreeing with Bata's father's earlier interpretation, it is unlikely that little children would be abroad in the dark or would be able to run away so quickly.

As the couple's interior village lies close to a forested area, the occurrence of ape-men so near a dwelling makes their account less questionable than reports of sightings inside or near other settlements. In fact, whatever they may have seen, their account seemed thoroughly naturalistic. Still, the couple considered the hominoids' visit inauspicious, and because Wula especially was fearful of what she'd observed, on the following day Bata chopped the ra'u tree down. After that, he added, they never saw the creatures again.

My main concern about the validity of this story is the discrepancy between Bata's initial suggestion that only his wife had seen the hominoids and her subsequent statement that not only her husband but her father-in-law, too, had observed them. Could it be that Bata was correct, and that Wula had included her husband, perhaps to validate her experience, and that, in addition, he was reluctant to contradict her in my presence? If Bata had contradicted his wife, however, it certainly wouldn't have been the first time I'd seen Lio spouses contradict one another! In fact, the next account comes from a woman who wanted to deny that her husband had seen what she'd seen. Considering the general reluctance to advertise encounters with lai ho'a, a more likely explanation is that Bata initially failed to mention his own observations because his wife had been the first to see the hominoids and, for this reason, he didn't want to provide a full account except in her presence.

The other ape-man sighting involving two spouses occurred in 2017, just three weeks before I heard about it. That year, when I returned to my usual lodgings in eastern Lio, my hosts were all astir about a villager who had recently experienced the loss of numerous chickens he kept in a locked pen. The fowls had not been removed but simply killed, apparently by a bite to the throat, and while the killer had sucked the blood, it had not eaten the flesh. The circumstances suggested the depredations of a familiar animal, most likely a palm civet.[4] However, when I met the disappointed owner—a matter-of-fact, no-nonsense sort of man named Gaja—he was certain that the culprit was a lai ho'a (ape-man). What's more, he said his younger brother's wife, Jata, had seen the thing.

Jata turned out to be a personable, outgoing, and relatively assertive woman in her forties. She confirmed that she'd observed the hominoid, one afternoon in early June just before sunset, around 6:00 P.M. The location was near rice fields bounded on one side by a stream and on the other by a secondary road that runs atop a high bank. This was where she and her husband cultivated and grazed livestock, and where her brother-in-law kept his chickens. Being near a paved road, the location is not especially isolated. Yet there are no permanent dwellings close by, and on the other side of the stream lies a forested hill, the former site of a long-abandoned village. A field-hut stands on the side adjoining the road, but this is used mainly to store chicken feed. No one sleeps there, and the site is usually unoccupied, especially at night.

Equipped with a bush-knife, Jata was cutting dried branches for firewood at the edge of a recently harvested field, just a few meters from the stream (see figure 6.3). She heard something moving about in a stand of trees on the other side of the stream. As is usual in such circumstances, she called "hoo" to signal her presence, to which something unseen responded with "hoo." Jata repeated the call, and it responded again, this time also throwing a stone that struck a stand of bamboo. Not long afterward, a small ape-man appeared, walking upright along a thick, low-hanging branch that grew across the stream. From the branch it jumped onto a large boulder before leaping onto the near bank, not far from where Jata was standing. The creature then walked past her, toward the middle of the rice field, where it stopped about 20 meters (66 feet) away.

As Jata stood watching the hominoid she called out to her husband, a man about the same age named Ari. At the time, Ari was tethering a cow somewhere near the field-hut located on the opposite side of the field. She shouted to him "what sort of animal is this?" and asked him to come and take a closer look at the "strange thing"—as she further labeled it. But Ari was reluctant. So Jata began walking, at a normal pace, toward the field-hut. When she moved in the creature's direction, it moved about the same distance, but it did not run away. When Jata arrived at the hut both she and her husband continued to observe the figure until it was visible no longer. Jata described it as seeming to disappear where it stood, though as she also mentioned, by this time the sun had set and it was dark.

While the couple were watching the hominoid, it "nodded" at them, lowering its head and bowing in their direction as though acknowledging that it could see them. When I spoke to him later, Ari independently mentioned the same detail. Specifically, he said that when he saw the thing it was bent over as though peering at something on the ground, but then it stood up straight and nodded its head. All the time they observed it, the creature made no sound.

Though Jata was briefly able to observe the hominoid at quite close range and longer than her husband, her description of its physical appearance was in some ways surprising. For one thing, she couldn't tell whether it was male or female, having noticed neither genitalia nor breasts—because, she said, it was getting dark. She was also unable to say whether it looked young or old. However, Jata confirmed that the creature stood completely erect and was not stooped. Despite references to specific facial features and what she described as a humanlike form, she was also unable to judge whether the ape-man looked more like a human or a monkey, describing its face only as "strange" or "odd-looking" and different from a monkey's. Just the same, she agreed that the face was unattractive (by human standards) and several times she spontaneously referred to the creature as an "animal."

At first Jata seemed unsure that what she'd seen was a lai ho'a (ape-man), pointing out that this was the first time she'd ever seen one. Later she said it must have been an ape-man as it had echoed her cry of "hoo." Standing erect on two legs like a human, the creature's height was little more than 60 centimeters (2 feet) and, as she mentioned spontaneously, it lacked a tail. Before I met Jata, her brother-in-law, Gaja, had also stated that she'd told him the figure had no tail. More generally he'd characterized it as a tailless bipedal monkey.

Implicitly referring to a human standard, Jata described the ape-man's legs as in proportion to the body, neither relatively long nor short. The feet were narrow or, at any rate, not noticeably wide, and though the ground was probably moist at the time, she saw no footprints. The body was thin or slight. Yet she responded positively when I asked whether it appeared muscular or sinewy. In contrast to otherwise dark skin, the chest appeared to be "pure white" (which, in Lio terms, could mean either white or very light). Previously, her brother-in-law had also mentioned the white chest, and the fact that Jata mentioned this detail both to him and to me indicates that she found it particularly noteworthy. White or light skin on the chest is atypical of Lio descriptions of ape-men. Two accounts of ape-man body hair, including one by Woda (a man introduced in chapter 2), mentioned light areas on otherwise dark hair or "white stripes" on either side of the chest. We might also recall the "white singlet" Mango described in his sighting report of a hominoid, summarized earlier. But no one ever specified variegated skin with lighter patches or streaks.

Another detail of Jata's description is equally unusual—her statement that the creature had no body hair. Two secondhand accounts mention glabrous hominoids, including Wonga's story of her deceased brother's encounter with another such being (in chapter 5), while other descriptions are silent about body hair. Jata also said she didn't notice any eyebrows. Funnily enough, this detail recalls Ura's story about a group of hominoids that lacked both body hair and eyebrows. Yet Ura's account is less credible than Jata's in numerous particulars and undoubtedly reflects a dream. Like the beings described by Ura, Jata's creature did have head hair, which was short, straight, and grayish or grayish brown. But this hair lay flat on the head, whereas the hair of Ura's dream assailants stood upright.

Although unable to make out the creature's sex or age, Jata was able to comment on other features of the face and head. The eyes were "bright" or "glittering, shining" (a description usually applied to human eyes), and the nose was short and flat. The eyes also appeared a bit sunken, perhaps indicating heavy brow ridges. The mouth was small and the lower face did not protrude—thus, as I was able to establish, resembling a human more than a monkey. Consistent with this, Jata described the chin as "short" and like a human's. (Curiously, she added that monkeys have "long" chins, whereas human chins become long only in old age.) Choosing between my specifications, Jata described the creature's forehead as low rather than high. Its ears were small and lay flat against the head. But I was unable to obtain any clear idea of the size of the head, which she simply described as small.

In response to my question, Jata thought the hominoid did not appear malicious or intent on harming anyone. After the sighting, neither she nor her husband had become ill or suffered other negative effects. Referring to the fact that the creature did not run away when it saw her but kept walking at a normal pace, she said it didn't seem frightened of her. And she claimed not to be afraid of it, pointing out that while she observed it she was carrying a bush-knife.

The account offered by Jata's husband, Ari, a less assertive and apparently less intelligent individual, differed little from his wife's. He described the hominoid as thin—no thicker or wider than his thigh—and as standing erect. As already mentioned, Ari also spoke of the creature as "nodding" in their direction. There is, however, a question of how much this man actually saw (or whether he saw the hominoid at all), as opposed to how much he heard from his wife. When I first spoke to her, Jata said only she had seen it, claiming that Ari had been afraid to leave their field-hut. We should also recall that Gaja, her brother-in-law and Ari's brother, spoke specifically of Jata as the observer, though this might only reflect the fact that she had seen it first (as in the case of Wula and her husband, Bata). Several times Jata mentioned that only she was not afraid of the creature. She further explained how it was specifically her individual "good fortune" (Indonesian "rejeki") that had enabled her to encounter the hominoid, and at one point she even suggested it wouldn't be possible for other people to see it. Contrary to this assertion, however, Jata later disclosed that her husband, Ari, had indeed seen the creature, "though only at a distance," and she reiterated that she had seen it first.

Despite the compelling nature of Jata's account and the fact that I recorded it just weeks after the event, there are reasons to question its veracity. Besides her prevarication over whether her husband also observed it, details of the subject appear atypical—including the hairless body, the "white" chest, and the apparent lack of sexually distinctive features (or Jata's failure to notice these). Since the sighting occurred just before sunset, there's the further question of how clearly she saw it. In this location at the beginning of June, the sun sets at 5:40 P.M., with civil twilight (the period during which, in clear weather, ordinary activities are still possible) ending at 6:02 P.M. The creature's apparent fearlessness may also raise a query, as may Jata's claim that she was not afraid.

Further doubt about the credibility of Jata's report arose when I again met the couple during my return to Flores in 2018. On my first visit that year, I wanted to speak to them about an entirely different matter, but when I arrived Jata immediately announced that, after my 2017 departure, she had seen the same creature again and in the same place. At first, she calculated that she'd seen it four times, but she later revised this to a single additional sighting. She added that her fifteen-year-old son (whom I was never able to interview separately) had also once seen it. Her husband, too, claimed he had observed the creature a second time, specifically in a tree from which it descended, returning to the other side of the stream after it spotted him. Evidently, the hominoid had not frightened him on this occasion, and referring to its first appearance he explained that he would have behaved more boldly had he not been recuperating from an illness at the time.

Perhaps more unsettling, however, Jata, who in 2017 had more than once described the hominoid as tailless, now described it as possessing a tail, though one that was very short—"only a little bit (of one)." Contrary to her former ambivalence regarding the thing's humanlike appearance, she further remarked that the creature "looked exactly like a human." Also, on this second encounter, Jata remarked that not only had the hominoid cried "hoo" before it appeared (also from the opposite side of the stream) but that she'd subsequently called out *wa'a ka* ("what's there?") and, on hearing this, the thing repeated her question. So on this occasion, but not on the first, the creature displayed a vocal habit regularly mentioned in popular descriptions of ape-men, a putative behavior with which Jata would certainly have been familiar.

Could it be that Jata elaborated her testimony not only by claiming a repeat visit but also by adding components of the popular image of ape-men? And might she have done this to further convince an investigator, whose persistent questions may have suggested skepticism, that she had actually seen an ape-man, and not something more ordinary? To Western readers this may seem a curious way to reduce doubt about the existence of an animal. But I would stress that properties of ape-men that seem fantastic to us are, for many Lio, just as "real" as features of a natural species.

If ape-men are real creatures, there is, of course, no reason why single specimens should not be seen more than once or by multiple observers. Possibly, the unusual physical details—the glabrous body and indistinct sexual features—reflected an immature or aberrant individual, perhaps one with a skin condition or suffering from partial albinism. However, given that Lio normally speak of ape-men as very rare and occurring in places far away from human settlements, Jata's story—including the claim that the creature returned at least once—seems exaggerated at best. We should also recall that she related her first sighting to her husband's brother, who'd recently suffered a loss of many chickens that he attributed to ape-men. Jata was convinced that what she saw was the culprit. So there's reason to believe that she'd expected to encounter a hominoid entering the field where her brother-in-law's (by that time empty) chicken pen was located.

Early in my visit to Lio territory in 2014, one of my hosts, an elderly man named Raga, mentioned a story he'd heard years before. Two fellow villagers had encountered a pair of ape-men. The older man, Ndori, had died in 1987, but the younger, Lalu, a man in his late sixties who was Ndori's brother-in-law, was still alive. So Raga introduced me to Lalu, who related their tale.

Over forty years previously, around 1969, Lalu and Ndori went to guard a plot that Ndori's father, Bedo, had planted with candlenut trees. It was the wet season, probably January, and it was drizzling at the time. The field was located in a forested area nearly an hour's walk from their village, and in the direction of Mount Bhéra, a 185-meter (600 foot) peak situated between the village and Flores's south coast. Although the slopes of Bhéra were cultivated in the early twentieth century, during the colonial period that followed, people began moving to lower elevations, and the top of the hill reverted to forest. In addition, the region upslope and southward of Bedo's candlenut plantation was declared protected forest in 1960, thus nearly a decade before Lalu and Ndori's reputed meeting with the ape-men. So there were no settlements in the vicinity in 1969, just as there are none today.

About 100 meters (330 feet) in length, the candlenut plantation extended from a lower-lying spot to a shallow rock shelter—little more than a limestone depression. This is named "Bedo's Cave" (Lia Bedo) because Ndori's father used to take cover there from the rain. Bedo had assigned Ndori to guard the plantation against the depredations of monkeys, which are still numerous in the area, and Lalu had accompanied him. As Lalu explained, monkeys will feed on or otherwise destroy the young shoots of candlenuts, and the plants, just a year or so old at the time, were still small and thus vulnerable to the primates.

Around four o'clock in the afternoon, the two men were approaching Bedo's Cave when they encountered a troop of monkeys in front of the rock shelter. A skilled archer, Ndori fired on the animals, causing them to scatter. But after the monkeys had fled, the men noticed two much larger ape-men squatting in the background. Although only about 15 meters (50 feet) away, the creatures did not move. According to Lalu, they were upwind of the men and so had yet to catch their scent. The two men had a dog with them. What role this played Lalu did not say. However, Raga, who had heard the story more than once before, later said that the dog did not immediately scare the hominoids away because they were at a higher elevation and so relatively safe.

Whatever the details, it wasn't long before the ape-men stood up straight, revealing a bipedal height of about one meter (3 feet, 3 inches). Ndori again fired his bow. The first arrow missed, swerving past one of the ape-men as it held up its arm. The creature—or perhaps the other creature—held up the other arm, and a second arrow whistled by on the other side. Raising the arms might suggest a reflexive attempt to deflect the arrows, but Lalu evidently favored a more mystical interpretation. Anyway, after having been fired on twice (and possibly with a dog in pursuit), the two creatures retreated, running bipedally by "lifting their legs in a sort of leaping fashion." They then scrambled on all fours up a bank to one side of Bedo's Cave and disappeared over the top.

Though the incident had occurred over forty years previously, Lalu's description of what he saw was notably detailed. There was a male and a female hominoid. The female appeared slightly shorter than the male and had breasts "larger than a female monkey's." Both creatures were covered in reddish hair, which Lalu compared to the hair of a nearby ginger-colored dog, but this covered only the limbs, buttocks, and backs. Their faces, bellies, and chests (and therefore the female's breasts) were bare, revealing skin about the same color as local humans'. The hair on the ape-men's heads was no longer than the body hair, and of the same reddish hue, but was thicker. The two sexes didn't differ in any of these respects. Nor did the hominoids have tails—something Lalu said he was sure of, because he'd seen their backs as they made their escape up the bank.

According to Lalu, the creatures' chests were "broad like a human's," though their overall build was neither particularly heavy nor slender. Their heads were large, with a width comparable to a small human's. And while Lalu described the faces as generally monkeylike, he said the noses were humanlike and the eyes large and prominent, "not sunken as in monkeys." Lalu was one of the few eyewitnesses who had much to say about ape-man feet. These specimens had "long" feet—"not as long as yours [the author's]," he added, but the length of a child's or young person's—and they were relatively wide. The toes, too, appeared long, though not as long as the fingers. But from top to bottom, the feet—specifically, the soles—were "thin" and, in this respect at least, rather "monkeylike."

Before his encounter, Lalu hadn't even heard the name "lai ho'a" (ape-man). But Ndori had. When he first saw the ape-men, Ndori exclaimed, "These are not monkeys, they are lai ho'a," and when they later told Bedo (Ndori's father) about the incident, the older man confirmed the identification, adding that the creatures probably lived in a higher-lying cave somewhere in the vicinity. Bedo was so alarmed by their story that he considered abandoning the location. For what it's worth, however, Lalu said he had not been afraid and that neither he nor Ndori had suffered any ill effect from their encounter.

When I first heard Lalu's story, I was skeptical because the details, remembered over a period of some forty-five years, were so precise. Yet Lalu had described his experience to others—including Raga, my host—and his memory may have been reinforced by regular retellings. Unfortunately, I was never able to question Lalu again about his experience. Shortly after our meeting, his wife suffered a major stroke that left her disabled and in constant need of her husband's attention. Because of her affliction Lalu became distraught and withdrawn, refusing to speak to other people and rarely leaving his house. Even three years later, when I ran into him accidentally, he claimed I'd mistaken him for someone else and refused to engage even in small talk.

In view of the unease the hominoids instill in many people, I've sometimes wondered whether Lalu linked his wife's decline to his telling me about his ape-man encounter. Nevertheless, Raga had heard the account more than once, and in the mid-1970s he'd heard it from Ndori—even though he and others described Ndori as a quiet man who spoke only when necessary. Anyway, Raga assured me that Ndori's account matched Lalu's 2014 version. So there's reason to believe the two brothers-in-law did see something, and more particularly something that looked hominoid. But if these were not ape-men, what were they?

Apparently, the only fantastical part of Lalu's story was the ape-men's ability to deflect arrows fired by a skilled archer. This recalls a more widespread idea, but one that concerns monkeys rather than ape-men, for Lio say monkeys will not only duck when arrows are fired but are also able to catch the missiles in flight and toss them to the ground or cast them back at the archer. (According to another claim, a monkey struck by an arrow can remove the shaft and somehow make the wound immediately disappear.) In fact, Raga later remarked that the ape-men avoiding arrows in Lalu's account sounded more like monkeys.

This could suggest that what Lalu and Ndori saw were no more than large monkeys. Lalu described the pair as staying behind after the other primates had all run away, a circumstance recalling the behavior of dominant male macaques, who will hold their ground when a troop is threatened from the rear. But a macaque troop usually includes a single dominant male whereas, in this case, there were two larger animals, and though an old study from the Flores region describes a troop of long-tailed macaques (*Macaca fascicularis*) led by two males, the strict dominance hierarchy in such troops would suggest this phenomenon is rare.[5] Besides, Lalu said the pair comprised a male and a female. He could, of course, have been mistaken. Yet the possibility of Lalu and his brother-in-law having observed nothing but a group of monkeys is at odds with other details of his account, especially the two creatures' far larger size, erect posture, bipedal movement, lack of tails, and other features suggesting something quite different from long-tailed macaques.

7

MORE REMARKABLE ENCOUNTERS

I f the last several reports are found compelling, then the following three cases, two of which concern observations of dead ape-men, should be found even more so.

Wolo's Story

Wolo (see figure 7.1) was born sometime between 1947 and 1949. I first met him in 2011 in his village in northeastern Lio. We'd been talking about lai ho'a (ape-men) for ten minutes or so when, out of the blue, Wolo mentioned a singular experience he'd had about forty years previously, while he was in his twenties, apparently in the early 1970s (most likely 1971 or 1972). We continued discussing details of his story when I next returned to Flores in 2014, and again in 2015.

As a young man, Wolo worked as a laborer, constructing modern roads and bridges, one of a number of projects that was sponsored by the Colombo Plan. Late one January afternoon he and three or four other members of a work crew were riding in the back of a truck. The truck was traveling on the Trans-Flores Highway, heading toward the south coastal town of Ende. Sometime between 4:00 and 5:00 P.M., the truck passed Ndu'a Ri'a, a traditional settlement whose name literally means "Big Forest." This is probably the highest point on the road, as it winds from the north coast to Ende on the south. Even during the driest part of the year, the weather can be damp and misty around Ndu'a Ri'a, as cloud descends from even higher mountaintops. But this was in the middle of the wet season. It had been raining and was still drizzling, and the road was shrouded in mist. During the 1970s most stretches of the main road were lined with thick forest. Ndu'a Ri'a villagers have now built houses and planted crops on either side of the road, but these changes began only in the following decade. And whereas market stalls selling local produce now line the highway in several stretches, in the 1970s there were only one or two irregularly tended stalls near the old village. Thus the truck with Wolo aboard was traveling a lonely, winding road, passing through jungle, and doing so in a mist.

At a location a short way past Ndu'a Ri'a, the driver, a Lio man named Raja, was rounding one of the many steep curves typical of most sections of the Trans-Flores Highway. Suddenly Wolo saw a hominoid figure descend from a bank at the left side of the road—Indonesians, it should be noted, drive on the left—and stop directly in front of the truck. The driver and other members of the crew saw it as well, but Wolo had a particularly good view as he was standing in the back of the truck, just behind the cab. Whoever saw it first, Raja was unable to brake in time, and he struck the creature, killing it instantly. The driver stopped and several men got out of the truck to inspect the victim, which Raja declared to be a lai ho'a (ape-man). Because Wolo was born and raised a few kilometers inland from Ende town, in a region where "lai ho'a" is unknown, he had never heard the name before. Nor was he familiar with the creature to which it referred.

Despite the incident having occurred decades ago, a particular value of Wolo's account is that it comes from a relative outsider, a non-Lio lacking any previous knowledge or prior conception of "ape-men." It also describes a sighting involving several other witnesses. Unfortunately, however, Raja—the driver—was long dead in 2011, and other members of the work crew had probably died as well. In any case, Wolo could not remember their names or details, and neither could surviving relatives of either Raja or Wolo, whom I was later able to question.

After inspecting the corpse, the men debated what they should do. According to a custom now observed throughout Indonesia, if motorists run over a cat they should stop, wrap the animal in their shirt or similar garment, and bury the carcass. If this is not done, it is believed that drivers will later meet with an accident. And this is what Raja wanted to do with the dead ape-man. From this we might infer that what the driver ran over was simply a cat. Yet Wolo was quite certain that the creature was not a cat and, anyway, the driver had identified it as a lai ho'a. There are other reasons why the men, and the driver especially, would have been inclined to bury the victim. It was a strange, extraordinary creature of a kind rarely seen. And, in several respects, it looked like a human being.

Among the tales discussed in chapter 4, one narrator similarly mentioned how a man, now long deceased, who found a dead ape-man caught in a trap, had wrapped the body in old clothes before burying it—in his words, "just as people do when they accidentally strike and kill a cat." But Wolo and his workmates did not bury their victim—even though they had tools available in the truck to do so. Younger members of the crew protested to the driver that it was too late in the afternoon, they were due back in Ende, and they were hungry. So, instead, the driver wrapped the body in an old shirt he had handy, and the men carried it to the edge of the road. Their plan was to bury the creature on the following day, when they were due to return that way.

Arriving in Ende town, one or more members of the crew—but neither the driver nor Wolo himself—related the incident to people there. When I spoke to him in 2014, Wolo mentioned that among the people his workmates told were two Australians, engineers or construction specialists stationed on Flores at the time. He also remembered their surnames as something like "Piper" and "Got." But in our final conversation in 2015, Wolo was unsure whether there were Westerners of any sort in Ende when the accident occurred.

The following day, the driver and Wolo did indeed return to the site of the accident, but neither the corpse nor the shirt was there. When I suggested the body might have been dragged off by a dog, Wolo objected, claiming that the spot was far from any habitation, so there could not have been any dogs in the area. People with whom I later discussed this thought Wolo's claim was unlikely. As they pointed out, village dogs often wander far from home and some become feral. But Wolo evidently favored a more mystical explanation, mentioning that the victim was a mysterious and frightening creature. (One word he used was the Indonesian "angker," which can mean "awful, sinister, dangerous," and of a place, "haunted, eerie.")

As regards the hominoid's physical appearance, what Wolo told me on different occasions was not always consistent—not surprisingly perhaps, as the incident had occurred some forty years previously. Apart from generally characterizing the creature as humanlike, when I asked him to describe it in 2011, he spontaneously stated that it had no tail. With his hand, he vaguely indicated a body length of 40–50 centimeters (1 foot, 4 inches to 1 foot, 8 inches), and in response to my question, he said it was female, as indicated by the genitalia. Wolo also described the body as covered in short hair like a cat's and light in color, and the skin as dark, like that of local people. Later, he described the hair color as "dark, like a monkey's hair." However, this discrepancy was resolved during subsequent conversations, when he explained that the hair was dark but appeared "light" as it was wet from the drizzle, and glistening. The creature's head hair was also short, and no longer than the body hair. The ears were humanlike, but the face rather resembled a monkey's.

Wolo repeated most of these details when I next met him in 2014. Although describing the creature as otherwise monkeylike, he again said the ears were wide and stuck out like a human's—a point he illustrated by holding his own ears outward. He also gave the same description of the body and head hair, though this time he added that the body hair was "not very dense." As in 2011, he spontaneously mentioned that the creature lacked a tail. However, while he had previously described the hominoid as female, three years later he claimed he'd never had the opportunity to observe the sex. Other details differed as well.

Whereas in 2011 Wolo had described the face as partly monkeylike, in 2014 he said the face was "human in form." Then, a year later, during my last meeting with him in 2015, he again described the face as resembling a monkey's. At the same time, he stressed that the face was not entirely like a monkey's, and it seems reasonable to conclude that such inconsistency reflects an impression of the creature's appearance as intermediate between a human and an ape.

In later years, Wolo's estimate of the creature's size was greater than the 40–50-centimeter body length he'd mentioned in 2011. In 2014 he compared the height to that of a nearby child, who measured 1.04 meters (3 feet, 5 inches). Perhaps his earlier estimate referred to the length of the body not including the limbs—as the creature lay prostrate in the road during most of the time that Wolo observed it. Anyway, he agreed that a length of over a meter would be far greater than any local monkey and added that the feet were like a human's, not a monkey's. All the same, when I asked what words should be used to distinguish sex in such a creature, he gave the Lio sex terms for animals (*mosa* and *metu*) rather than those employed for humans.

I met Wolo for the last time in 2015, when we continued discussing his 1970s sighting. Our conversation produced few further physical details. He did, however, indicate a height for the creature shorter than his 2014 estimate, as this time he compared its height to that of another local child, who stood about 80 centimeters tall (2 feet, 8 inches). This was also the first occasion that Wolo mentioned the hominoid having genital hair. This, he agreed, would suggest that the specimen had been an adult—though, as in the previous year, he couldn't say whether it was female or male—this time explaining that the genitalia were obscured by the hair.

Toward the end of our 2015 conversation, Wolo made a more serious departure. One of his workmates, he said, had suggested the victim might simply be a monkey. This led me to ask whether it had a tail, and Wolo responded that "there was a tail but it was short" (that is, shorter than a monkey's). This contrasted with the long tails of local macaques, and Wolo also confirmed that the creature's face was humanlike—though, answering another question, he said the ears were monkeylike. (In earlier conversations, of course, he'd portrayed the ears as human.) These new details were surprising, not least because on all previous occasions Wolo had spontaneously described the victim as tailless. One might dismiss this as a misunderstanding—the old man seemed tired at the time, and he may have thought I was referring to popular representations of ape-men, with which he would have become familiar during the forty-odd years following the accident. Alternatively, it is conceivable that, on this occasion, Wolo depicted the creature as more simian than previously in order to counter any suggestion, perhaps implied by my repeated queries about its physical appearance, that it had somehow been human after all—and that he had, therefore, been involved in the killing of a human that had gone unreported.

Later in the same meeting Wolo also suggested—speculatively, it seemed—that the hominoid had crossed the road on all fours. In all previous conversations he'd explicitly described the creature as erect and bipedal and, as just noted, during this same conversation he'd described it as standing as tall as a child of 80 centimeters. More specifically, in the notes on our first meeting in 2011, I'd written that the hominoid "definitely moved bipedally, like a human, not on all fours" and, further, that Wolo "was quite certain of this." Also, during a subsequent interview, when he confirmed the bipedal locomotion, Wolo spontaneously added that, in this respect and in its diminutive size, the creature was "like a small child."

This further turnabout might be explained by the same factors accounting for what might be called the "sudden growth of a tail." As for whether it walked on two or four limbs, it is unclear how long Wolo, the driver, or anyone else observed the creature moving before the truck struck. Still, the contradictions remain puzzling. Partly because it was getting late, I didn't confront Wolo with his previous statements about bipedalism and taillessness. These were issues I thought I might broach during my next visit to Flores, when I planned to take up an offer Wolo had made to accompany me to the spot where the hominoid was killed, which he claimed he would still be able to identify. This, however, never came to be, for when I returned to Flores in 2016, I found the old man terminally ill and unable to speak.

To fully appreciate Wolo's story, I should say more about his character. Wolo was a man with no education beyond primary schooling. When I first met him in 2011, he was a widower who had long been living with one of his sons in a village on Flores's north coast. During our second meeting, in 2014, I noticed that since I'd first met him, Wolo had dyed his hair black, as many Florenese men do after they turn gray (though, interestingly enough, not the women). All the same, he seemed older than his years—under sixty-five in 2011—and while he appeared alert and generally competent, he seemed older and slower than on our first meeting—an apparent change that could further illuminate the inconsistency in his description of the hominoid during our final meeting a year later.

Before we met in 2011 Wolo had been recommended to me by a fellow villager whom I'd met in a nearby market. This person directed me to Wolo as a man possibly able to provide information on a rumor I'd heard about someone, a year or two previously, capturing an ape-man and bringing it to Wolo's village. Wolo rejected the rumor as arrant nonsense. More generally, he revealed himself to be a serious but affable man, soft-spoken and taciturn, though quite prepared to tell me what he knew about ape-men—an openness he maintained throughout our acquaintance. It may have been beneficial that, when I first met him, I was traveling with a young Indonesian Catholic priest. Yet Wolo did not know the priest, and I've no reason to believe the cleric's presence affected the substance of what he told me. In any event, as a Catholic himself, it's unlikely that Wolo would have deliberately told mistruths.

Simply stated, both then and in successive years, Wolo struck me as a straightforward and honest—one might even say guileless—man, not given to exaggeration or having any axe to grind. Certainly, if his story had been told by some other Lio I've known, I would have been far more skeptical. While forthcoming in his statements and always willing to answer questions, it soon became clear that Wolo was fuzzy on dates and people's ages. He was even uncertain about his own year of birth, though such uncertainty is not unusual among Lio of middle age or above. In establishing his age and the date of the road accident, however, I was eventually able to obtain more reliable information from members of his family and from relatives of Raja, the driver (who died around 1978).

Another thing I discovered from conversations with relatives and village-mates of both Wolo and Raja—including Raja's widow and Wolo's co-resident son—was that neither man had ever told any of them about the accident or its hominoid victim. Wolo confirmed this, adding that he never told his wife either. The only explanation he offered was that he didn't think anyone would be interested. This surprised me at the time. Yet both Raja's widow (who died in 2015) and his surviving siblings separately described the driver as a quiet man who spoke only when necessary (thus like Ndori, described in connection with Lalu's story in the previous chapter), and Wolo struck me as similar.

Unless Wolo's story is a complete fabrication, there are just three ways to explain it: the truck struck a known animal (a monkey, cat, or dog); it struck a human, presumably a child; or it struck something corresponding to usual Lio descriptions of ape-men. If it had been a familiar animal, however, surely this would have been immediately obvious to the driver and crew. Both domestic and wild animals are run down all the time on Flores roads, so the accident would have been in no way unusual—and therefore unlikely to be remembered by anyone decades later. Anyway, if it had been a monkey, as Wolo seemed to suggest on our last meeting, or a dog, the driver would not have been compelled to bury it. Also, as Lio regularly consume monkeys and dogs, the men, who were hungry at the time, would almost certainly have kept the carcass, or even cooked and eaten it on the spot. In fact, in 2014 Wolo reported the driver as sarcastically suggesting to the crew (who, as mentioned, were complaining of being hungry) that if they thought the victim was just an ordinary animal, they should burn off the hair and roast it!

In reference to possible monkeylike features (including the newly added tail and quadrupedal locomotion), Wolo ended our final 2015 conversation with the vague speculation that "maybe the thing had joined together with monkeys and thus had already become a monkey"—a statement vaguely reminiscent of ideas concerning the origin of ape-men discussed earlier. All the same, Wolo was adamant that the creature could not have been an ordinary human being, and for this and a variety of other reasons, the second possibility—that it was simply a human—is equally unlikely. If the victim had been a child, the driver or one of the other men would either have reported it or kept very quiet, especially when they arrived in Ende, the administrative center for the region and the location of police headquarters. So this leaves only the third possibility.

Tegu's Discovery

Like Wolo's story, the following account concerns a dead specimen, in fact two dead specimens. What's more, the bodies were found at widely separate times by the roadside near the village of Ndu'a Ri'a, in the same vicinity where Wolo's accident had occurred.

In 2015 I was lodging temporarily in another highland village, following up reports of villagers who claimed to possess ape-man relics. Among these was a man whose daughter-in-law, an outgoing and rather garrulous young woman named Wea, derived from Ndu'a Ri'a. After I'd finished speaking with her father-in-law, Wea approached me with several stories about ape-men. Although they varied in credibility, one especially caught my attention. It concerned a Ndu'a Ri'a man named Tegu, who a few years previously had been alerted to the presence of a dead ape-man on his land. The corpse, she said, had first been discovered by a number of people returning from church, herself included. After finding the corpse, Tegu and a friend whose Christian name was Fanus (short for Stefanus) buried it somewhere. However, as Wea further claimed, later that evening the creature's relatives disinterred the body and took it away.

Although I'm still unsure whether Wea actually saw the creature, she mentioned several quite specific physical details. Found lying facedown, the creature's head was "almost the same as a human's" and the body was covered in fairly sparse, light-grayish hair. The face resembled a monkey's, and the nose was "like a skull," which she explained to mean covered in scabs or mange. Information subsequently given by Tegu himself included neither of these details. Nevertheless, Wea described the location of Tegu's house, and I was easily able to find it the next day. Tegu was not home, but his wife was. When I mentioned how I'd heard that her husband had found a dead creature of a strange kind some years previously, she seemed not to know what I was talking about. But I left my cell phone number and asked her to have her husband contact me.

Since similar requests for people to call me had previously gone unheeded, and because Tegu didn't know me from Adam, I was rather surprised to receive his call early the following morning. Tegu, it turned out, was a man of fifty years who held a bachelor's degree in agriculture from Bandung in Java, and who cultivated his own land mostly to provide vegetables for sale in the market at Ndu'a Ri'a. Friendly and enthusiastic in answering my inquiries, he was clearly intelligent, though his knowledge of indigenous culture appeared limited—perhaps not surprisingly for a man who had spent years away from Flores. I had several phone conversations with Tegu that year, but it was not until my subsequent visit the following year (2016) that I was able to meet him in person. What he told me on all occasions was remarkably consistent.

Some years previously, probably in 2010 and on a Pentecost Monday (May 24th, if 2010 was the year), Tegu and others had attended church. Afterward he returned home and was taking a nap when his wife and her sister—a woman from the Ende region who had been working as a nurse in Ndu'a Ri'a—awakened him. In a state of alarm, the women reported they had found a corpse or carcass (both English words are covered by a single local term) and urged Tegu to go and inspect it. The spot was on the other side of the road from Tegu's house and a short distance to the west, atop a bank and near a path Ndu'a Ri'a villagers used to walk down to the main road, including when they traveled to and from the church.

What Tegu found there, lying in undergrowth and not far from a large tree, was the corpse of what appeared to be an elderly female hominoid with a "human face," whose naked body was covered in short, fine hair. It was lying facedown with legs bent in a kneeling position and arms drawn toward the chest, thus in what is commonly called the child's pose. Interestingly, I had heard elsewhere that this is the pose adopted by sleeping ape-men. Another source had described it as a position assumed by both ape-men and monkeys when they die. And Tegu suggested that the hominoid, after traveling from a place unknown, had become exhausted and had fallen asleep on the spot and subsequently expired.

Tegu did not identify the creature as a "lai ho'a" (ape-man) but as an "earth spirit" (tana watu)—contrary to the usual representation of these beings as entirely supernatural and invisible (chapter 3). Lai ho'a, he said, were "bad things," whereas he expressed sympathy for the elderly hominoid. Feeling obligated because the pathetic creature had died on his land, Tegu quickly wrapped the body in a woman's garment. With the assistance of a friend—the man named Fanus—he hastily constructed a plywood coffin and put the corpse inside. Tying the box to his motorbike, Tegu later traveled, alone, to a spot on Flores's south coast, where he deposited the makeshift coffin in the water and let it drift out to sea. Contrary to what Wea had told me, he assured me that he did not bury the body.

At the time of the discovery, Tegu said, his wife had suggested that the hominoid may have been savaged by dogs. But he immediately rejected this, pointing out that he saw no wounds or blood on the corpse. He also denied that it could have been struck by a vehicle, as the body was found too far from the roadside, on top of a raised bank. Tegu provided more details of the creature's physical appearance during our meeting in 2016. Though he viewed the body only briefly—because, as he noted, he wanted to dispose of it quickly—the face resembled that of a "small woman" with a "well-formed" (Indonesian "mancung") nose. It could not, he confirmed, have possibly been a monkey, but he was equally sure it couldn't have been a human. (Indeed, if he'd thought it was human, his course of action would certainly have been very different.) Looking extremely old, the creature's head and body hair were "white (or light-colored)," while the skin was dark, like local people's, or more specifically "dark brownish, like the skin of an elderly person." Tegu further described the body hair as "fine but quite dense," and as "dense as a puppy dog's." (Wea, by contrast, had characterized the hair as rather sparse.)

The deceased's head hair was straight, thus not curly or kinky like the hair of many Lio, and was longer than the body hair. At one point Tegu spoke of the hair as growing to the shoulders, but I could get no more exact indication of the length. During our 2015 conversations Tegu mentioned the length of the head hair as the sole reason he thought the dead hominoid was female. Hair length has been an indicator of sex difference in Lio for less than a century, since the advent of Western-style short haircuts for men. However, in 2016, Tegu said he knew it was female from the breasts, which resembled those of an old woman. But, curiously, he said he never noticed whether the chest was hairy.

The hands and fingers were "very small," as were the feet, which were arranged with the soles turned upward (as in the previously mentioned child's pose). Recalling a more specific detail, Tegu added that the small toe on one foot was bent over the next toe. In response to my question, he suggested that the hominoid would have stood about 65 centimeters tall (2 feet, 2 inches). However, he also estimated the length of the body from the head to the base of the spine as 50 centimeters (20 inches). So unless the legs were inordinately short, the creature would have been much taller, perhaps approaching a meter.

As always when investigating reputed sightings, I asked whether the deceased hominoid had a tail. Tegu said it did, a very short one less than 4 centimeters (1.6 inches). But he added that the tail was covered in hair, like the rest of the body, so it is unclear whether he actually saw it. Similarly, at one point Tegu suggested that the hominoid had four rather than five digits on each hand (though apparently a full complement on each foot), whereas later he revealed that the fingers on both hands had been clenched into fists and that he'd never actually been able to see the digits. Indeed, earlier he had described the arms as being held against the chest. As mentioned previously, digits numbering more or less than five is a feature Lio occasionally attribute to some fully supernatural beings as well as to ape-men. So it may well be that in this respect, as with regard to the hidden tail, Tegu—who, as noted, interpreted the corpse as belonging to an "earth spirit"—was drawing on popular images of spirits or ape-men.

Before leaving Tegu's house in 2016, I also spoke to his wife. Named Keo, she hailed from the Ende district, specifically from Ndao, which lies immediately southwest of the Lio region. Our interview was not particularly successful. She was shy, nervous, and reluctant to speak, and I was unable to question her when her husband was not in earshot. Keo confirmed that she had come across the body with her sister, but I couldn't clarify who had spotted it first. At one point Tegu had spoken as though this had been his wife's sister. As she repeated several times, the experience had severely frightened both women, so they hurried away. When I asked for details of the body, Keo said she couldn't say, as they had not gotten close enough. She added that it "is not usual to see such things, so I was afraid."

It seems, then, that the women did see the corpse, if only very briefly and from a distance. As Keo volunteered, they had first become alerted to its presence by the smell of something rotting, which they thought might be a dead pig or dog. At this point Tegu interjected that he never detected any smell, adding that bodies only begin to smell two or three nights after death—a remark suggesting that he thought the creature had died more recently. He did, however, say there was a cow tied up nearby, straining at its tether as though frightened and trying to get away. Perhaps illuminating the disagreement over the smell, the tree near where Tegu found the body was a species of fig (*Ficus variegata*; Lio *ara*). In Southeast Asia, orangutans and gibbons, as well as humans, regularly eat the tree's sweet, edible fruit, but when they fall to the ground and rot, they give off a very disagreeable smell. Possibly, then, it was this odor that the two women mistook for a rotting carcass. Or, alternatively, Tegu did detect a smell but attributed this not to the dead hominoid but to rotting figs.

For reasons I explain in a moment, I never got to meet Keo's sister. Nor was I able to locate Fanus, the man who'd helped Tegu construct the coffin; he lived elsewhere and, it seems, just happened to be visiting at the time. In any case, Tegu thought Fanus, whom he knew only by his Christian name, had probably not seen much of the body, as Tegu was alone when he wrapped it up, thus concealing its face and other features.

When I departed from Tegu's house in 2016, I was thus left with several questions. Tegu seemed open to speaking again. But, in fact, this was to be our last meeting. About a week later, I phoned Tegu, but he was far from home on some business. He said he didn't know when we could next meet, and—contrasting noticeably with his previous attitude—he seemed unenthusiastic about the prospect. Subsequent phone calls went unanswered. Then, after another week, I happened to be passing through Ndu'a Ri'a and made a stop at Tegu's house. Only his wife was there. She was visibly distressed at my appearance, stating that she and her husband were busy and had no time to talk to me. They'd already told me everything I wanted to know, she added. And when I said I still wished to talk to her sister, she refused to provide information on where exactly the woman (who had long since moved back to Ende) was living or how I might contact her. Needless to say, I was taken aback by this reception—as were two Lio men I was traveling with. And compounding my surprise was the fact that during our meeting in his house, Tegu had telephoned his wife's sister to ask what she remembered of the dead ape-man, but she was unavailable at the time. Having thought about it over the last several years, I'm now quite sure that resistance to my meeting Tegu again stemmed mainly from his wife, and that this was because she remained fearful about what she and her husband had discovered and therefore did not wish to discuss it any further.

Other questions remain concerning Tegu's story. First, in 2016 he claimed he'd told no one else of the discovery, so the only other people who knew of it were his wife, her sister, and his visiting friend. In 2015 he'd mentioned how "people had told him that if he had taken the creature's teeth or head hair, he could have obtained [supernatural] power." Tegu did not credit this, but the remark obviously suggests that he had indeed told other people. All the same, inquiries I made in 2016 among other people in the vicinity disclosed no one who had heard Tegu's story.

Wea, the woman who first directed me to Tegu, claimed she and others had come across the body—or at least detected an odor—as they were returning from church. If Wea did not actually see it, then details of her account—including how the body was lying face down, the smell, and other particulars of its appearance—indicate that she had heard a detailed report from somebody. Because she'd mentioned Fanus in her account, one possibility is that this had been Tegu's friend. Tegu himself seemed unclear about who Wea even was. Yet since he explained that he'd felt privileged to have found the deceased creature, and in view of the general idea that only certain people are ever able to encounter such things, my impression was that Tegu wanted to represent not only his experience but also himself as exceptional, even though he didn't completely deny that other people (including his wife and her sister) may also have seen it. In the same vein, he added that it was only because of his discovery that he, a Lio cultivator, had come to meet me, a foreign professor, and that this was all God's will.

Another peculiarity of the story lies in Tegu's claim that, rather than burying the body (as Wea stated), he disposed of it in the sea, and at a spot some distance from his highland home. This was Nanga Nesa, a beach near Ende town. What particular significance this place might have had for Tegu I was unable to discover. But the spot was well known to him; he traveled to Ende regularly; and, anyway, it's likely he wanted to dispose of the corpse as far from his home as possible. When I asked him why he had taken the body to the sea, he said his parents and grandparents had told him that beings like the hominoid, which he called "things of the mountains," should be "taken to the sea [when they die] because there are no humans there." Though he didn't expressly say so, it seemed clear that Tegu thought it unsafe to bury such a creature on dry land, especially in proximity to human settlements—an idea perhaps echoed in Wea's claim that, after the corpse had been buried, the deceased hominoid's relatives had dug it up and taken it away. The reputed episode is also reminiscent of a notion I'd heard more than once before, that ape-men bury their dead on beaches because the sand allows them to dig without tools.

But 2010 was not the only time Tegu had reputedly encountered a dead hominoid. When I first spoke to him in 2015, he mentioned that such creatures are extremely rare, suggesting that "of a million people, only one person will ever see one." Nevertheless, during our first conversation and subsequently, Tegu claimed that he'd come across another corpse, another female, when he was seven years old—thus (as he was born in 1965) around 1972. It lay beside a roadside stream a short distance from his present house and thus quite close to the location of his 2010 discovery. At the time, though, the road was narrow and unpaved; either side was dense jungle; and there were no habitations nearby. He'd told no one else about this earlier discovery—and most definitely not his parents, as they would have been angry that he had frequented such a lonely and "dangerous" place, a location he described as a haunt of spirits.

The earlier specimen was, in several respects, different from the more recent one. Tegu was sure the creature, with an estimated body length of 40 to 50 centimeters (1 foot, 4 inches to 1 foot, 8 inches), was of the same type. But it appeared young, like the hominoid equivalent of an adolescent girl, with small, firm breasts. Lying on its back, its long, straight black hair was spread outward, and partly because of what Tegu described as an "attractive" face, it looked like a "doll" (i.e., a manufactured child's plaything). Yet Tegu denied it could have been a doll, remarking that such things were not available on Flores at the time. Also, the naked body, which he viewed from a distance of about 3 meters (10 feet), was covered in short, fine hair, while the skin was relatively light, "like that of a Chinese." Responding to my question, Tegu thought it was not impossible that the hominoid was unconscious or asleep rather than dead. But, overtaken by fear at his sudden discovery, he quickly ran away, not staying long enough to inspect the body closely.

Given the widely acknowledged rarity of ape-men, the chances of the same person seeing two specimens, both of which appeared female and both lying apparently dead in what was virtually the same location, raises further questions about Tegu's reports. Yet the naturalistic details of both accounts, as well as the circumstances in which I came to meet the man and the matter-of-fact way in which he related his experiences, make what he told me, especially regarding his 2010 discovery, compelling. Were it not for the body hair and small size, the earlier corpse could conceivably have belonged to a human adolescent. All the same, there is no reason to believe Tegu's description does not reflect something zoological if not human or hominoid. And his story reveals another intriguing coincidence. The year Tegu encountered the younger creature, around 1972, was about the time that Wolo's truck struck and killed the hominoid whose corpse subsequently disappeared. What's more, the accident apparently occurred very close to where Tegu found his body, if not at the exact same spot. So I should add that Tegu and Wolo never knew one another and never lived anywhere near one another.

Noko, Dhiki, and Bara: A Man, His Cousin, and Her Daughter

The final report relates to a creature seen much earlier than Tegu's more recent find. It is further distinguished by the fact that it was observed by two people I was able to question at length and on several separate occasions. One was Noko, a friendly yet serious man and the very model of an unsophisticated Lio cultivator. Noko had never attended school (at least not regularly) and was quite unsure of his own age, although I subsequently learned he had been born—or at least baptized—in 1966. The other witness was Noko's female cousin, Dhiki, a similarly unsophisticated, uneducated, and illiterate woman some years older than Noko, who suffered from what was apparently a nervous disorder—a condition recognized by both Noko and Dhiki's village-mates. Yet another witness was Dhiki's daughter, about ten years old at the time. Unfortunately, though, I was unable to question this woman as, some years previously, she had moved with her husband to Indonesian Borneo.

In 2015 I was detained in a highland village in central Lio. Wisely, my motorcycle chauffeur wanted to travel alone over the steep rocky trail leading to our destination, a neighboring settlement, to ensure the people we were looking for were there. As it happened, they were not. Anyway, it was during my wait that I came across Noko and got talking to him about local knowledge of mystery hominoids. I asked him whether he'd heard of "an erect, bipedal, rather humanlike creature, hairy and in some other ways resembling a monkey but larger than a monkey." Noko answered straightaway that he had, that the creatures were called "lai ho'a," and furthermore that he had once seen one. He then revealed that the specimen had been simultaneously observed by his female cousin and her young daughter.

The sighting occurred in the late 1990s, probably toward the end of the wet season, around May. Early one morning Noko was in his garden, a plot located near the edge of a forest and quite recently cleared, perhaps 3 kilometers from his village (see figure 7.2). There were no permanent settlements nearby. Next to Noko's garden, however, was another, worked by his cousin Dhiki and her husband. But on this morning, only Dhiki and her daughter, Bara, were present. While her mother picked vegetables, Bara wandered some distance away, toward the other end of the garden. At this end, Dhiki's plot, like Noko's, was bounded by a dry streambed, and just beyond the streambed there were large boulders and several felled tree trunks and branches.

All of a sudden, a dog that had followed Dhiki from her village ran toward this end of the garden and started barking. Alerted by the noise, Bara, who was close to the dog, turned to see what it was barking at. Alarmed by what she saw, the little girl immediately cried out to Noko (who was closer than her mother), telling him to come quickly as she had seen a strange "child." More particularly (and perhaps rhetorically), she asked in Lio "whose (or "what sort of") child is this?" (*ana sai gha?*). Both Noko and her mother then ran quickly to her daughter's aid. What all three of them saw, at a distance of perhaps 10 to 15 meters (30 to 50 feet), was a hominoid, sitting or squatting among the felled tree trunks and throwing stones at the dog. Where the creature had come from neither Dhiki nor Noko could say. But both speculated it may have lived in either of two caves located near the gardens, and Dhiki further suggested it may have come out into the open to warm itself in the morning sun.

Everyone was extremely frightened by what they saw—Noko recalled how he began sweating profusely—and for this reason they did not observe the creature for long. Noko called off the dog, initially getting its attention by pelting it with a stone, and they all quickly fled homeward, with the two females screaming and crying in distress. That evening, the little girl developed a high fever and her hair began to fall out, growing back only after a month or more. Both Noko and Dhiki told a number of people of the incident but, according to Noko, no one was bold enough to venture near the spot for a long time afterward. Neither the creature nor anything like it was ever seen again, either by the eyewitnesses or by fellow villagers.

When I met them, Noko and Dhiki were living in different villages. I questioned them about the incident separately on two occasions in 2015, and again in 2016. I have no evidence that the cousins had any opportunity to coordinate their reports. Nor, indeed, was there any reason they should have wanted to do so. What they described was similar, as regards both the circumstances of the encounter and the physical appearance of the creature, and where they varied, their accounts were mostly complementary rather than contradictory.

Independently, the two eyewitnesses described a hominoid about the size of a child of six or seven years. Although it remained sitting in the same spot throughout the encounter, it appeared bipedal and generally humanlike. In fact, Noko consistently represented it as some sort of "human being," though not a "human like us" but something essentially different. For one thing, the naked body was covered in hair. Had it stood up, Noko said, its full height would have been either about 85 centimeters or one meter (2 feet, 9 inches or 3 feet, 3 inches). This he inferred by comparing the creature to two young village boys, the taller of whom had a sitting height of over 50 centimeters (20 inches). Similarly, in 2015 and 2016, Dhiki compared the creature's height to those of different local children I was able to measure. One stood 86.5 centimeters (2 feet, 10 inches), and the other 1.02 meters (3 feet, 4 inches). In both instances, estimates of the head size corresponded with the estimates of height. Thus Noko thought this was about the same size as the head of a young boy who stood 1.27 meters (4 feet, 2 inches; see figure 2.1), while Dhiki compared the head to that of a nearby girl of six or more years (see figure 7.3).

According to both observers, therefore, the hominoid was much larger than a local macaque. Responding to my question, neither Noko nor Dhiki said they had seen a tail. How old the thing might have been neither could say. But both thought it was mature, with Noko basing his interpretation on the gaunt appearance of the face and rather sunken cheeks. At the same time,

Noko mentioned that because the head and body hair was dark rather than "white" or "light," it did not appear elderly. Remarking how it appeared to be female—or an *ata fai* (the Lio phrase for a human female)—Dhiki claimed she'd seen small breasts and nipples. Noko too initially identified it as female, but what distinguished the creature's sex he could not say, and when we spoke again in 2016, he seemed uncertain about the sex.

The body hair was thicker than a human's but sparse in comparison to a monkey's. Both Noko and Dhiki initially described the color of the body and head hair as brown or reddish brown, whereas later, in 2016, they both gave the color as "black" or "dark." While Noko simply characterized the body as hairy, Dhiki was more specific, describing only the chest and belly as free of hair. (When questioned again later, Noko thought there was no difference between the chest hair and hair on the rest of the body.)

How the head hair differed from the body hair was unclear. At one point, Noko said it was not particularly long, though somewhat longer than the body hair. In the subsequent year, he estimated the length as perhaps 15 centimeters (6 inches), adding that it appeared disheveled and untidy. Dhiki spoke of a "medium" length and, with her hand, indicated that it grew to the shoulders. Noko could not confirm this because, he said, the creature consistently faced toward him. Like the dead hominoid found by Tegu, both eyewitnesses said the hair was straight—that is, not kinky or curly—and that it fell over the forehead. Dhiki seemed to describe the hair as entirely covering the forehead and even the eyes. But it later transpired that she had referred to an occasion when the hominoid bowed its head in the direction of the woman and her daughter, after one or both screamed upon seeing it. Noko similarly stated that the only motion the creature made (besides throwing stones at the dog) was when it craned its neck forward, and he imitated it, bending over as if to bow. Remarkably, this detail recalls the action attributed to the hominoid observed by Jata, and perhaps her husband as well, in another place and on a far more recent occasion.

Responding to my question, Dhiki said the creature's skin was "black" or "dark" like that of local people. She also compared its complexion to that of a particular girl, whose skin, however, struck me as possibly lighter than the local norm. Noko, by contrast, seemed to characterize the hominoid's color as darker than local people. But as he also mentioned, the skin looked dirty, and he further characterized it as "not in good condition," possibly meaning rough or blemished.

The two observers gave quite similar accounts of the creature's face. Dhiki described a "long" face that appeared intermediate between a human's and a monkey's. Noko characterized it as humanlike and about the size of a small child's but "ugly, unattractive" and "not good." According to Dhiki, the cheeks were "fat" or "full" like those of people, thus contrasting to the sunken cheeks of a monkey. This may seem to contradict Noko's aforementioned description of a gaunt face and sunken cheeks, but in this connection Noko was contrasting adults and children rather than humans and monkeys. Answering my question, Noko had also described the arms as "thin," but this was the only specific mention by either eyewitness of the hominoid's limbs or general build.

Both Noko and Dhiki indicated that the forehead was "small," sloped backward, and partly covered in hair. Dhiki also spoke of a relatively "prominent" forehead, but as with the cheeks, this implied a contrast with macaques rather than a complete resemblance to modern humans. As for other facial features, only Dhiki described the mouth or lower face as "protruding," like a monkey's, whereas Noko simply characterized it as humanlike. Both described the nose as like a human's—more prominent than a monkey's and, in Dhiki's estimation, not flat but "well-formed" ("mancung," the same word used by Tegu for the dead hominoid he found). Neither could comment on the teeth. As both remarked, they never saw the thing open its mouth, and it never made any vocal sound. Dhiki had rather more to say than Noko about the ears and eyes. Whereas Noko described these simply as humanlike but smaller than an adult human's, Dhiki spoke of ears that were relatively large and stuck out, like the ears of a human and unlike those of a monkey. Also, during our second interview in 2016, she remarked that the outer ears appeared to be turned or folded forward—a condition which, in humans, Lio call *bongo*, meaning something like a cauliflower ear.

Dhiki seemed to have been particularly drawn to the eyes. In 2015 she'd spontaneously remarked that these were "clear" like a human's (and thus unlike a monkey's). In 2016 she added that they appeared large and wide, features she said made her especially afraid of the creature. Also, later in our conversation, she described the eyes as "opened wide"—a phrase that, in humans, describes a sign of surprise, shock, or fear. If accurate, this detail could suggest that the hominoid was as scared of its human observers as they were of it, or became so after the females began screaming. Both in 2015 and 2016 Dhiki offered other details of the creature's eyes. The eyebrow hair was dense and long—so long, she later claimed, that it seemed to partly cover the hominoid's eyes. The lashes, too, may have been long, though she was rather unclear on this point. When I later questioned him, Noko was unable to confirm these particulars, mentioning that, after seeing the creature, he had not stuck around long enough to observe such details.

Yet another feature mentioned by Dhiki but not by Noko was the belly, which she spontaneously described not only as hairless but noticeably distended. Indeed, several circumstances suggest that, while Dhiki's young daughter had first spotted the hominoid, Dhiki too may have seen it before Noko and so was able to observe it a bit longer. Even so, Dhiki, who especially during our second meeting in 2016 seemed slightly distracted and unclear in her thoughts, was less consistent in her descriptions. For example, whereas she'd earlier described either the creature's feet or legs as "long," she later stated that, like Noko, she had not seen the lower limbs at all, as the hominoid remained sitting among rocks or vegetation. Also in 2016, Dhiki referred to long fingernails—a feature far more often mentioned in popular accounts. And whereas Noko consistently claimed there had been one dog present at the encounter, in 2015 Dhiki said there had been no dog, while in 2016 she claimed there were two dogs.

More startlingly, Dhiki first stated that Noko had not been present, only her daughter and herself. But later she denied this as well. Especially as a fellow villager of Noko's who'd heard the story also thought Noko had not been there, at one point I began to doubt Noko's account. But details he provided of the circumstances of the encounter, including his description of his psychological response to the sighting, as well as his apparent character and general demeanor (including his reaction when I suggested that perhaps only Dhiki and her daughter had seen the hominoid) weigh against this claim as, of course, does the woman's subsequent revision. Had Noko not been present at the encounter, then he could only have heard about it from either Dhiki or her daughter. However, in that case, it is unlikely that either would have mentioned to Noko such specific details of the hominoid's appearance or behavior as he was later able to relate to me.

As noted, Noko claimed that what they saw was some sort of human, albeit one that was small and hairy-bodied and thus qualitatively different from humans like himself. Despite several monkeylike features mentioned by both eyewitnesses, they strenuously denied that it could have been a monkey, with both remarking that they were extremely familiar with monkeys. The far larger body and head size of the hominoid, of course, further count against the monkey hypothesis. When he first saw it, Noko said, he immediately knew it was what local people call "lai ho'a," and Dhiki too said she knew this name and what it referred to before the encounter.

If the thing were neither a monkey nor an ape-man, the only other possibility would seem to be a wayward, misshapen, and perhaps mentally abnormal human child. But this appears ruled out by several physical features, not least of which is the hairy body, and also by the fact that there were no settlements nearby. Even if there had been, had such a child been living in the vicinity this would almost certainly have been known to Noko and Dhiki. Or if they were not familiar with the child at the time, the youngster would have been identified as the object of their sighting after the fact. (Lio are familiar with Down syndrome and, in fact, one youngster so affected was living in Dhiki's home village when I spoke to her.)

Throughout our conversations, Noko made it clear that he regarded the creature as ill-intentioned. During a brief meeting in 2017, he further described it with an Indonesian phrase translating as "a ghost (or spirit) person who is not good" ("orang hantu yang tidak bagus"). This might suggest that he conceived the creature to be a supernatural being. But if it were somehow a spirit—beings Lio describe as essentially invisible—then one would have to explain how three people witnessed it simultaneously. The hominoid's apparently malevolent character explained for both Noko and Dhiki why Dhiki's daughter, the first to spot it, had subsequently become seriously ill with a fever and suffered sudden hair loss. Loss of head hair—in children, more specifically the condition called telogen effluvium—can be caused by severe stress or emotional shock, as can fever.[1] The young girl's mental distress was likely compounded by the reactions of fear and panic exhibited by her mother and Noko after she drew their attention to the strange being. Nevertheless, if either of her symptoms were the result of the ape-man encounter, they would attest to the subject having been an anomalous—humanlike but nonhuman—creature of a sort that none of the witnesses had ever seen before.

Despite a few discrepancies (all attributable to the fact that the sighting took place nearly two decades previously), both Dhiki's and Noko's accounts are coherent and realistic. Since ape-men are widely described as afraid of dogs, one might question the claim that the creature did not run away when the dog barked, holding its ground and even pelting the advancing canine with stones. Yet a similar detail is found in Lalu's story (recounted in chapter 6), where a barking dog, while scaring off a number of accompanying monkeys, also failed to put a pair of ape-men to flight. In addition, two other stories reviewed in the last chapter, related by Malo and Jata, concern hominoids that did not immediately flee in the presence of a human. That the hominoid threw stones at the barking dog may also raise a question. However, Jata described the hominoid she saw as throwing a stone (though with what intention she couldn't say). Also, several Lio I asked considered this behavior credible, and some mentioned witnessing monkeys, too, defensively throwing stones at people or dogs.

Among the most detailed accounts I recorded, the subjects of Wolo's story, Tegu's 2010 observations, and Noko's and Dhiki's reports appear to have been specimens of a single zoological kind. Despite certain inconsistencies, and the fact that Wolo's account, by far the oldest, is shorter on physical details than the others, all these subjects were of about the same form and size. Except of course for the immature specimen found by Tegu, all seem to have approached the upper size limit—between 80 centimeters (2 feet, 8 inches) and a meter or more—revealed in Lio descriptions of ape-men in general. All were also hairy-bodied, or at least more hirsute than local humans.

Although described in varying degrees as resembling humans—recall the "well-formed" nose mentioned by both Tegu and Dhiki—all specimens seem to have been somewhat simian in their facial appearance. This is less clear from Tegu's account. Yet, especially as Flores Islanders sometimes describe very old people as looking like monkeys, his impression that the more recently found corpse belonged to an elderly creature possibly reflected a simian appearance rather than a specimen that was actually old. Of course, the same man described the younger creature he claimed to have seen as "attractive, cute." But the term he used (Indonesian "cantik") can apply to nonhuman animals as well as people, especially where young specimens are concerned.

Further in agreement are the observers' reactions to what they saw. Both Noko and Dhiki emphasized how their experiences frightened them, and though fear was only implicit in Tegu's account, his wife certainly admitted to having been very afraid. Wolo revealed little of his personal feelings on encountering the accidentally killed hominoid. This fits with his generally quiet and undemonstrative character and perhaps also the forty years that had elapsed since the incident. Nonetheless, he similarly described the creature as "frightening."

Relevant to the credibility of these three reports are the circumstances in which I met the eyewitnesses. All three were encountered opportunistically, which is to say quite by accident. Although other people directed me to two—Tegu and Wolo—not even Tegu appeared to have broadcasted his experience widely. In fact, Wea and any others who knew of it probably heard of it not from Tegu or his wife, but from either the wife's sister or Fanus. Assessing the comparative credibility of the three cases is difficult, not least because credibility reflects a variety of factors (the circumstances, location, and date of the putative sighting; the character of the witnesses). However, all things considered, Noko and Dhiki's encounter would seem the most believable. Although the amount of detail in Dhiki's account is perhaps surprising in view of the time that had elapsed since the sighting, the veracity of her testimony is nonetheless supported by substantial agreement between her account and Noko's. Anyway, all three sightings are compelling. Indeed, in comparison with other eyewitness accounts, they provide the best evidence for the existence of a scientifically undiscovered creature Lio call "lai ho'a."

8

THE LITTLE NAKED MAN AND
OTHER EXTRATERRITORIAL SIGHTINGS

n this chapter I make a brief detour. Readers may be asking: If Lio have observed ape-men in recent decades, why haven't people in other parts of Flores seen them? Actually, more than one story suggests they have.

Two Ngadha Men in Nagé Country

In 1975, two young men, then in their early twenties, were hiking around the Ebu Lobo volcano in the Nagé region (see figure 8.1), about a hundred miles to the west of Lio, when they saw a small naked "man." The individual fled at their approach and was never seen again. I first heard the story from one of the observers in 1999. But the full significance of their report only became apparent sixteen years later, in 2015, when I was able to interview the second eyewitness. By this time I had also recorded the majority of Lio accounts of local ape-men, including those discussed in the last chapter.

This case deserves our attention for several reasons. For one thing it concerns the experience of two rather unlikely eyewitnesses who had observed the subject simultaneously. What's more, the two observers were not local people but young men, born in 1952 and 1953, who derived from the ethnolinguistically distinct Ngadha region, more specifically, from Ngadha villages well to the west of Nagé territory. This is important. Nagé, we should recall, is home to a tradition of small-bodied hairy hominoids invariably described as having become extinct hundreds of years ago, after being exterminated by Nagé ancestors. Named "ebu gogo," the hominoids inhabited the higher slopes of the Ebu Lobo volcano, and it was on the lower slopes of the mountain that the two Ngadha men saw a small-bodied hominoid. At the time, the men knew nothing of this tradition. So they had no reason to link what they saw with the legendary Nagé hominoids, nor did they have any expectation of encountering such a thing.

The observers' backgrounds are also important. When I first met him in 1999, one man, Felis, was a civil servant in Bajawa (the capital of the Ngadha administrative district) and the former head of a high school. An educated man, Felis is (and considers himself to be) a very modern person, little given to mystical ideas and disinterested in and generally unknowledgeable about local culture and indigenous beliefs. Although apparently more conversant in such matters, the other man, Hilar, received the same education as Felis, his classmate throughout secondary school, and served as a teacher and administrator at a high school in Kupang (the provincial capital, on Timor Island) until his retirement in 2015.

Equally important is the fact that I was able to speak to both men completely independently of one another. When he spoke to me in 1999, and again in 2005 and 2015, Felis had lost contact with Hilar. He also assured me he had not spoken to Hilar about their experience at any time during our acquaintance—though the two men could, of course, have discussed what they saw just after seeing it. In 2015 Felis's son was able to get me an email address for Hilar, and when I told Felis that I'd arranged to meet his old schoolmate during a coming visit to Kupang, he asked me to ask for Hilar's telephone number (Felis's usual method of communicating), which he no longer had. Later, when I met Hilar, he too confirmed that he hadn't spoken to Felis for years—and least of all about their shared hominoid encounter. So there's no reason to believe they ever coordinated their descriptions of what they saw. Yet while their accounts differed in several details (as did Felis's on the different occasions when I spoke to him), what they told me was remarkably similar, even on quite specific points.

I first heard about the sighting in 1999 while visiting Felis's Nagé father-in-law, a longtime informant, with whom I happened to be discussing the Nagé hominoids named ebu gogo. Overhearing our conversation Felis, who was also visiting his parents-in-law and whom I'd never met previously, mentioned that, years before, he and a friend had been traveling in the region when they spotted a small naked human figure, with longish head hair, on the lower slopes of Ebu Lobo. The only other feature Felis mentioned was that the man's lower arms appeared somewhat "bent" or "crooked." This, he suggested, might have indicated an advanced age. (It might also recall the flexed forearms some Lio attribute to ape-men.)

I wasn't able to speak further to Felis in 1999 and didn't find the opportunity to do so again until 2005. The following summarizes what he told me in 2005 as well as on two subsequent occasions.

Felis and Hilar's sighting occurred in September 1975, not long after the two men had completed their formal education. To celebrate this achievement, they'd planned to hike around the volcano, beginning in the Nagé village of Wudu and proceeding to the southeast. This was about full moon and, after spending the previous day with a friend in Wudu, the youths set out in the late afternoon, around four o'clock; as this was the dry season and the weather was hot, they had chosen to travel at night. Around 4:30 P.M. the two men were following a path some distance upslope from Wudu village. They were passing some gardens, some of which had been burned off in preparation for planting in the rainy season (usually beginning in October). Then suddenly, to their right, they noticed what both described as a small naked "man"—an apparently human male—in one of the plots.

I'll continue to refer to the being as a "man," as it looked human and both eyewitnesses were sure it was male. The word Felis used was Indonesian "orang," a term that means "human" but in other contexts can refer to other things, most notably in "orang hutan," an Indonesian name for the orangutan (the ape *Pongo pygmaeus*). Neither Felis nor Hilar could suggest what else their humanlike subject might have been. Still, at one point, Felis said he could not be sure whether it was human or some other kind of "creature" (Indonesian "makluk," a reference to animals and spirits as well as humans).

When the strange little man saw the youths, he quickly ran away, entered an area of vegetation, and "disappeared." Felis never mentioned how long they observed the individual, but Hilar thought it was no more than a minute. The men did not bring along dogs, as villagers often do when they travel, nor did they carry any weapons. Though frightened by what they saw, this didn't prevent the young travelers from stopping to inspect the spot where they first saw the figure. There they found peanuts dug up and whole nuts and shells scattered about, suggesting that the little man had been consuming the nuts.

Both eyewitnesses described the figure as standing about one meter tall, though on one occasion Felis manually indicated a height which appeared somewhere between 1.2 and 1.3 meters (around or over 4 feet). The subject appeared quite sturdily built and muscular, but his belly was somewhat distended. Felis, especially, thought the face looked somewhat "old." But as he noted, the color of the head hair was black, not gray, as one would expect in an older person, and the ease with which the little hominoid ran away also suggested a younger individual.

The subject had body hair, probably like a normal human adult, according to Hilar, but he did not appear particularly hairy. However, as both men pointed out, they could not see clearly enough to make an exact determination. The man's skin was dark, even a little darker than local people—perhaps suggesting that any body hair would not have been particularly noticeable—and Felis further described the subject's body as covered in dirt. At no point did the individual make any sound. Responding to my suggestion that the little man could have been bathing or was about to bathe—something that would explain his nakedness—both informants disagreed, pointing out that there was no water source nearby.

When I met Hilar in 2015, I found him a friendly and accommodating man. Initially, he seemed a little reticent, evidently puzzled by the reason for my request to meet him (which I hadn't previously mentioned). But he became more relaxed when he realized I spoke Indonesian, and that I'd known Felis and his family for many years. After explaining my interests in Florenese culture, history, and languages, I remarked how I'd heard that, in their younger days, he and Felis had once hiked around the Ebu Lobo volcano. He remembered the occasion clearly and offered details. I then mentioned how Felis had said that, on the trip, they had encountered a "strange human" (Indonesian "manusia ganjil"), a point he immediately confirmed. Hilar's description of the man's height, build, skin color, and head and body hair corresponded closely to what Felis had told me: the figure stood erect, was about a meter tall, and had a distended belly, and the complexion and body hair differed little from local people's. Particularly striking was Hilar's mention of such minor details as the protruding belly and the peanuts (which he identified straightaway when I asked whether there was any evidence the man had been eating anything in the garden). In addition, when I asked whether anything about the individual's arms appeared peculiar—recalling Felis's reference to bent or crooked lower arms—Hilar responded that both lower arms appeared "fat" or "thick" relative to the rest of the limbs and could have been "swollen." At the same time, both men described the length of the arms, and also the legs, as in proportion to the rest of the body—that is, neither noticeably short nor long. The head, too, was in proportion.

As should be expected, the two men differed on some details. According to Hilar's description the head hair, which he specified as shoulder length, seemed not quite so long as Felis had suggested. Hilar also described the little man's face as "funny-looking" and reminiscent of things seen in films from Java. When I asked for clarification, he immediately responded that he was talking about "kurcaci," one name for the mischievous foreign elves depicted in television programs. As we already know, Lio people tend to identify these imaginary figures with their local ape-men. But it's important to note that in 1975, the year of Hilar's observations, films or TV shows depicting such elves were barely available on Flores, so Hilar's assessment was almost certainly a reconstruction based on later experience of these media. From what he said I somehow got the impression that the face may have looked something like a "troll" doll, with distorted or cartoonish features. In a similar vein, Hilar suggested the figure was "childlike," and that he and Felis found it "amusing." Nevertheless, he said the little man appeared to be an adult, though he could not estimate the age. When I later asked Felis he was unable to confirm any of these details, though he reiterated that the man somehow looked "old." Yet he retracted an earlier suggestion, that the man appeared "stooped," affirming that he could stand upright, like an ordinary human.

Hilar further remarked that when he and his companion first saw the little man, he was "running about" (Indonesian "lari lari") as though "wanting to play." This, too, Felis denied. However, when Felis described how the figure fled when it saw them, he used the Indonesian phrases "lompat lari" ("jump [and] run [away]") and "lompat-lompat" (meaning "to jump or skip around"). As Indonesian lompat ("jump") does not refer to "quickly getting up (e.g., from a sitting position)"—as in English "jumping up from one's seat"—the description might suggest that the hominoid in fact ran, either while fleeing or prior to this, with a sort of jumping or bounding gait. So Hilar and Felis's descriptions of the figure's movements could partly refer to the same thing. Anyway, the reference to "jumping" remains intriguing in relation to Lio descriptions of the peculiar gait sometimes attributed to ape-men.

Further disagreement concerns the precise course of events from the time the young men first saw the little man until his eventual disappearance. As both men agreed, however, the subject appeared startled when he first saw them and ran away quickly—revealing that (in Hilar's words) he was "energetic" and "agile," and further suggesting he could not have been elderly. It was also during the little man's flight, Felis recalled, that he fully realized the individual was naked, particularly mentioning the sight of his bare buttocks as he ran away. The observers also disagreed about how far they were from the little man when they first saw him. Hilar was quite sure it was about 7 meters (23 feet), a distance he estimated by comparing it to the distance from where we were talking to another spot. By contrast, both before and after I interviewed Hilar, Felis said the distance was 30 to 40 meters (100 to 130 feet). But when I raised this with Hilar, he denied it, sticking with 7 meters. Felis's estimate, he suggested, may have referred to when the little man was running away. Otherwise, I am unable to explain the difference.

A discrepancy between Felis's earlier and later accounts concerns footprints the subject left behind. In 2005, when the topic first came up, Felis said they'd seen no footprints when they went to inspect the ground trodden by the little man. In 2015, Hilar stated they had found prints, which were small but in proportion to the subject's overall size. Then, in the following year (2016), while Felis and I were discussing how the individual had run away, Felis referred spontaneously to having found footprints, also mentioning that these appeared human. A skeptic might suspect the informants conferred in the meantime. But I've already given reasons for not thinking so. And if they coordinated their accounts on this single point, we must ask why they evidently didn't do so on others. More likely, in the course of our conversations spanning some sixteen years, Felis recalled something he'd not remembered earlier.

There's no reason to believe the young travelers did not see something. But what was it? One possibility is that the little man was a local person who was in some way abnormal. When I asked, neither witness said the face was "monkeylike" or appeared "ugly"—despite my impression of a "troll-like" visage and Hilar's reference to media elves, which often are indeed ugly. So for this and several other reasons, there's absolutely no indication that what the men saw was a monkey or other local animal.

Other than the subject's small size, the only possible physical abnormality was the peculiar appearance of the lower limbs. The distended belly seems less relevant as this could have reflected dietary deficiency or disease—though it's interesting that the same feature was reported in the last chapter for a Lio hominoid. (As we'll later see, it also fits *Homo floresiensis*.) As I was able to establish from questioning about the limbs and the head, all of which were in proportion to the torso, the tiny figure could not have been an achondroplastic dwarf, a condition with which both informants were familiar. Anyway, there's no reason such an individual, who by all indications was an adult, should be naked and should run away at the sight of the two men, even though they were strangers. And the same goes for a proportionate dwarf (or "midget"). On this score, I suggested that the subject might have been deranged. Both eyewitnesses agreed this was possible, though neither seemed entirely convinced. (I raised the possibility because, several times in my travels on Flores and neighboring islands, I'd come across adults described as "mad" or "crazy" who were in the habit of taking off their clothes.)

After making inquiries in Wudu and nearby villages, I was unable to find support for any of these hypotheses. No one I asked who'd been an adolescent or adult in the 1970s had ever heard of a short, deranged person who would regularly or occasionally go naked. A partial exception was a man from Nanga, a village several kilometers to the east of Wudu, who'd recently died and whose relatives described him as stripping naked when someone upset him. But he wasn't short; indeed, he was rather tall by local standards, and moreover, he didn't show signs of mental instability until the early 1990s.

Whatever his mental condition, a man this small—little more than a meter in height—would surely have been remembered by someone in the vicinity. The only other possibility might be an odd-looking or abnormal human child, for example one suffering from progeria or who, for some other reason, looked adult, or even quite old. But the individual's build argues against this, as does the fact that no child of this description was remembered either.

What's so striking about the subject of Hilar and Felis's sighting is, of course, its resemblance to the Lio ape-men. True, the individual's body did not appear especially hairy. But then neither did creatures seen by several Lio. Also, while neither of the Ngadha men thought the individual they saw had an ugly or simian face, by their own confession neither was able to view it for long, as they mostly observed the little man's back as he ran away. This seems to be the reason neither man noticed the genitals either, even though both consistently described their subject as male. All in all, the figure appeared to correspond most closely to the hominoid seen by the Lio cousins, Noko and Dhiki. Recall that Dhiki's account, too, mentioned a distended belly. If it were indeed the same thing, this would not be particularly surprising. Although Lio is more mountainous than Nagé, there's no reason such a being could survive in the first region but not in the second.

A More Recent Report from Nagé

Though unusual, what the two Ngadha saw in Nagé territory is not quite unique. Investigating an entirely different matter in 2015, I met Lowa, a quiet but obliging middle-aged Nagé cultivator who described something he'd seen while alone in a forested area in 2010 or 2011. This looked and behaved very much like Hilar and Felis's subject. The main difference was that it was smaller—around 65 centimeters (2 feet, 2 inches) according to Lowa's estimate—and though he described it as a "child," it had pubic and body hair. While the head hair was long, it was also frizzy or curly. But as Lowa mentioned spontaneously, the naked hominoid had a distended belly, as did the little man seen by Felis and Hilar.

Lowa identified his subject as a "kurcaci," again employing the Indo-nesian name for the elves of modern media. Why Nagé people do not identify small, fully visible, and evidently humanlike beings with their ebu gogo hominoids is another question. Nagé often depict the ebu gogo as extremely hairy whereas the little men seen by Hilar, Felis, and Lowa were not. But probably a more decisive reason is that the legendary Nagé hominoids are deemed extinct. This may be an instance of things going unnoticed or not making an impression because they're no longer sup-posed to exist. Because it equally applies to scientists who dismiss certain undocumented or "cryptic" animals as imaginary, we'll return to this issue in the final chapter.

A Western Observer on a Neighboring Island

For a final example of extraterritorial hominoid sightings, we leave Flores and travel westward to Sumbawa, the next large island in the Lesser Sunda chain. In eastern Sumbawa, the Bimanese, who comprise the majority population, speak of small hominoids with very short tails that inhabit the Wawo highlands, in the southeastern part of the island. According to anthropologist Michael (Mike) Hitchcock, now a retired professor at the University of London, the Bimanese seem not to distinguish these creatures entirely from the Wawo people themselves, the human inhabitants of the Wawo mountains, whom they consider culturally inferior, less than fully human, and (as mentioned in chapter 2) as possessing small tails.[1] Given the similarity to the Lio ape-men, readers can be forgiven for thinking this may be a simple case of cultural continuity, or people of related cultures holding similar beliefs. In fact, the Bimanese, a large group long converted to Islam, are quite different from the Lio—in regard to religion, social and political organization, general culture, and history.

Far more noteworthy, however, Professor Hitchcock himself reports seeing a group of relatively large bipedal primates apparently corresponding to these hominoids. His encounter dates back to October 1982. Mike was twenty-seven at the time and carrying out doctoral research on Bimanese material culture, but he also made visits to the Wawo region, at the time a very isolated area. As Mike recorded in notebooks he showed me in 2005, one morning he was climbing alone to the top of a hill in search of a carved stone he'd heard might be found there. Eventually, the narrow path he was following became, in his own words, a "tunnel of foliage" about a meter high, so he had to crouch and eventually crawl. Though at first Mike assumed that villagers had fashioned the tunnel by pushing back foliage, to provide access to the stone, it soon became clear that no modern human could possibly go through it walking upright. But the thick vegetation wasn't the only impediment, for just a few meters away he was suddenly confronted by a group of three or four "angry monkeys." Shrieking at him, the creatures were clearly not going to allow him to pass. So, very sensibly, Mike backed off and returned to the village where he'd been staying.

Because of the low-hanging vegetation, Mike only got a dim view of the creatures and he judiciously noted how easy it would be to make mistakes in these conditions. But as he told me when I first spoke to him in 2005, he was sure of several details. To begin with, the creatures all stood upright (or "bolt upright" as he phrased it in 2021) and at no point did they go down on four limbs. The largest of the group stood "nearly one meter" tall, the same height as the tunnel. At no point did he see any tails—a circumstance perhaps explained by the fact that the creatures were facing him throughout the encounter. As he further remarked, they appeared to be "bobbing up and down" on two legs, and though he couldn't be sure, they may have been slowly moving toward him. More recently (in December 2020), Mike mentioned spontaneously how these hominoids did not have long head hair, adding that "he would have noticed if they had, as he could see their heads." But the light was too dim to make out hair on the body.

Though in his notes Mike describes the hominoids as "monkeys," as he told me in 2005, these details have "always troubled him," and for this reason the incident has remained on his mind for nearly forty years. (Professor Hitchcock, I should mention, studied primatology as part of his undergraduate anthropology degree.) The creatures' vocalizations sounded lower-pitched than those of long-tailed macaques, the only monkeys known to occur on Sumbawa, as on Flores. More importantly, they didn't look like macaques; according to Mike, there was something "odd" or "unfamiliar" about the creatures, and "this experience was different from [his] many previous encounters with monkeys in Indonesia."[2]

It's important to note that neither in 1982 nor subsequently has Mike's research (which focuses on the Bimanese rather than the Wawo people) been concerned with primates, mystery hominoids, or other animals. For this reason he never related his experience to the Wawo people, and he's unable to confirm whether Wawo know of or speak about large, bipedal, and apparently tailless primates like the ones he saw. It goes without saying that, in all particulars

Mike was able to record, the creatures sound very much like the Lio ape-men. To a lesser degree, so do the small, short-tailed hominoids that the Bimanese say live in the Wawo mountains. I am not suggesting that because Mike is a Westerner and a senior academic to boot, his testimony is any more valid than that of Lio people who claim to have observed local ape-men—though I am, of course, extremely grateful for his candor and his allowing me to write about his experience. Nevertheless, it surely says something about the possible reality of small, bipedal eastern Indonesian hominoids that people of very different cultural backgrounds can report seeing what sound like the very same things. In fact, as noted previously, another non-Indonesian researcher told me of a small humanlike creature they saw while traveling alone at night on Flores. (As it happens, this encounter too was reported from the 1980s.)

PART IV

WHAT TO BELIEVE

WHAT DO STORIES OF APE-MEN TELL US AND WHAT ELSE COULD THEY BE?

Returning to the Lio stories, what can we learn from all eyewitness reports taken together? And how do they compare to popular descriptions of ape-men and to secondhand accounts?

Why Some Eyewitness Reports are More Credible than Others

In chapters 6 and 7 I ordered sighting reports intuitively, beginning with those that struck me as least credible and proceeding to the most. Now that we've reviewed the reports individually and done some unpacking, the order appears justified according to specific criteria. Credibility, of course, means the extent to which an account appears to reflect a natural creature (which is mortal, comes in two sexes, appears in different sizes, and so on) and doesn't correspond to any familiar animal, as well as the location and time of a sighting. Other circumstantial factors can be equally important. For example, if ape-men are as rare as Lio say, reports by people who claim to have encountered them more than once raise a question. The same goes for stories told by mystical practitioners, who can enhance their reputations by claiming contact with unusual beings or an ability to see things that other people have not—or supposedly cannot. So it's significant that none of the people whose experiences were described in chapter 7 and chapter 8 were mystical practitioners. As we saw in chapter 7, Tegu described encountering hominoids on two occasions nearly forty years apart, yet his more recent sighting, especially, appeared credible in all other respects.

On these grounds I'm inclined to dismiss at least half of the stories summarized in chapter 6. For a start, the least plausible sightings strongly suggest encounters with more familiar creatures, especially monkeys. This conclusion doesn't follow only from their smaller sizes, head hair continuous with body hair, possession of long tails, and the fact that subjects were reported as moving, at least part of the time, on four limbs. It also finds support in the fact that most sightings occurred in places near the coast. Ape-men, of course, are generally described as creatures of high mountain forests that normally maintain a distance from permanent human settlements. So it seems significant that stories of sightings that contained evidently fantastical elements—for example, ape-men being killed or buried on a beach or encountered inside a house or village—were also set on or near the coast.

This is not to say that all highland sightings reflected plausible encounters with ape-men. One such sighting, shared by Koli and Watu, involved a specimen that was most definitely a monkey—not least because, apart from its reputedly shorter tail, the two young men described it as looking exactly like a monkey! A similar case concerned two putative sightings of long-tailed monkeylike creatures by a man named Kesu, also during his younger years. Unlike the two youngsters, however, Kesu generally struck me as a tall-talker, and indeed, his stories incorporated the mythological theme of cave-burning as well as a hominoid subject—who, unusually, had a proper name—visiting his uncle in a dream. (Even so, Kesu told me about his first encounter in 2005 and again in 2016, and what he related on the second occasion, though more elaborate, was not fundamentally different.)

As for reports that cannot be put down to monkeys, most concerned sightings either after nightfall or in a darkened cave, where visibility and thus details of the subject were limited. To be sure, over half of these described hominoids observed in motorcycle headlamps or other artificial light. Yet none is as impressive as reports of creatures seen in daylight—in regard to the quality of physical details or circumstantial factors. These daytime sightings include the experiences related by Wolo and Tegu (and possibly Wea and Keo) and by Noko and Dhiki (chapter 7), as well as the detailed accounts given by Lalu and Jata (discussed at the end of chapter 6). Besides, these last several sightings took place in the wet season—though in two instances, toward the end of that season—which by general agreement is when people are more likely to encounter ape-men. And we should also not forget that the eyewitnesses saw specimens that were dead, though owing to circumstances—including their emotional reactions to what they found—none of them saw the bodies for long or inspected them closely.

Yet another factor enhances the plausibility of these reports. In every instance more than one person observed the hominoid subject or subjects simultaneously—as did Wula and her husband Bata (chapter 6), whose sighting took place at night. And where two or more observers were available for questioning (Noko and Dhiki; Tegu, Keo, and Wea; Bata and Wula; and Jata and Ari), the witnesses gave similar accounts of what they saw—though in one or two instances, doubt remains about whether one person did, in fact, see what another saw.

Of course, not all stories by people claiming they'd seen ape-men were free of apparently imaginary elements. Among the most fantastical instances, Ura's tale surely reflected a dream. So, in all likelihood, did Bajo's account of a strange-looking creature he saw from his house after waking in the middle of the night. The setting, moreover, was a coastal village, as was the location of another encounter—inside his house, no less—related by Rina, a man suffering mental impairment. That said, none of the other eyewitnesses reported empirically implausible details. No one claimed to have seen what looked like an ape-man flying or transforming into another creature. And only in Bajo's recollection of a probable dream—and, of course, in stories told by people who evidently encountered monkeys or civets—did anyone report actually seeing a tail.

Still, there were two accounts, both related by elderly mystical practitioners, that are difficult to accept on other grounds. One was Rawi's story of killing an ape-man on a south coast beach. The other was Gali's tale, suspect in several respects, about his encounter with hominoids burying one of their kind in a similar coastal location. But not all reports from individuals whose character or mystical reputation made them less than ideal witnesses were completely incredible. Though I've characterized him as an otherwise untrustworthy source, Pora described an encounter with a group of hominoids that, in its details and circumstances, sounded little different from what was related to me by Malo, a far more credible witness. Also, some features of Pora's subjects call to mind the small female hominoid with long flowing hair that Tegu, one of my most convincing informants, remembered coming across in the 1970s. The more general point, of course, is that even habitual deceivers will tell the truth sometimes, and even the most outlandish tale can have some basis in reality.

Rawi, the ape-man killer who later carried his victim's skull on a trip to New Guinea, is another case in point. After making this claim, the elderly mystic went on to give detailed descriptions of hominoids he said he'd observed at an inland location, which in most respects sounded thoroughly naturalistic. Comparing all accounts I recorded, though, Rawi remains a complete wild card, and I still have difficulty placing what the old man told me in relation to other Lio testimony.

Eyewitness Reports and Popular Accounts

Both in quantity and substance, details of accounts from people who describe seeing ape-men obviously vary. Yet taken as a whole, eyewitness reports reveal important differences from descriptions by non-eyewitnesses (the popular accounts summarized in chapter 2). Specifically, reports by eyewitnesses—and here I necessarily exclude observers who apparently reported seeing monkeys—paint a picture that is decidedly more human-like than do popular depictions—which, by contrast, portray the hominoids as more monkeylike, even while also mentioning erect posture and bipedal locomotion.

Estimates by eyewitnesses of hominoid height actually don't differ much from those given in other accounts. Popular accounts included heights slightly taller than those reported by eyewitnesses. The tallest subject of an eyewitness account was 1.03 meters (3 feet, 5 inches), while the greatest height mentioned by a non-eyewitness was 1.2 meters (3 feet, 11 inches)—in each instance based on comparison with the height of a nearby child. The difference could partly reflect age and sexual differences among creatures seen by eyewitnesses, whereas popular accounts are presumably based on some notional average height. More importantly, though, just one observer mentioned a height below 60 or 70 centimeters (2 feet or 2 feet, 4 inches), and tellingly, the informant was Gali (chapter 6), who reported a dead creature just 30 centimeters (1 foot) long, and whose report was easily the least plausible of all I recorded. A few popular descriptions also mentioned heights this small, but as previously suggested, these most likely reflect influence from modern media images.

When it comes to the hominoids' hair and facial features, differences between specimens reputedly seen and the image of ape-men conveyed in popular descriptions are more pronounced. Both eyewitness and popular accounts described ape-man faces as either monkeylike or humanlike. But with popular accounts, there was a nearly even split between the two alternatives. In contrast, a far higher proportion of eyewitnesses (5 of 7) specified the subject's face as humanlike. (Here we might especially recall Noko's statement, that lai ho'a [ape-men] are humans but "not humans like us.") An equal number of eyewitnesses and non-eyewitnesses—four of each—described ape-men as both humanlike and monkeylike, but in different contexts or at different points in our conversations. We should also remember that many eyewitnesses were unable to observe their subjects closely, mostly due to poor light.

The difference between observers' and popular accounts is confirmed when we look at specific features of the face, including the eyes, ears, nose, and teeth. Whereas non-eyewitnesses hardly mentioned these, five eyewitnesses described eyes that were "clear" or "bright, glittering" or in other ways like a human's, and another observer spoke of eyes that were "truly human." Observers who encountered dead specimens whose eyes were shut could not, of course, comment on the eyes. Details provided by four of five eyewitnesses who described a specimen's nose suggested it was humanlike or more like a human's than a local macaque's—for example, "well-formed" and "more prominent than a monkey's." Describing a monkeylike lower face combined with a human-like upper face, another observer specified a nose that was "flat." So did the sole non-eyewitness who mentioned the nose. But then Lio also characterize some of their fellows as flat-nosed.

Details of the ears, forehead, or chin were absent from popular accounts. By contrast four eyewitnesses characterized the ears as humanlike, even though their descriptions were neither particularly detailed nor, in one case (Wolo), consistent. Just two eyewitnesses mentioned the chin, which one described as "curved" (Rawi) and the other (Jata) as "short" and thus "humanlike." At the same time, these two observers described the forehead as, respectively, "small" and "low," seemingly suggesting a more simian appearance.

Descriptions of the teeth are less decisive in regard to human or monkey comparisons. Only Rawi spoke explicitly of humanlike teeth, but how far this reflected any of the encounters he claimed remains unclear. Two other eyewitnesses (including Laka, discussed in chapter 5) referred only to canine teeth, which were prominent and, in one instance, sharp—as did two non-eyewitnesses. For the rest, two popular accounts simply mentioned monkeylike teeth, which could, of course, refer to large canines. Three other non-eyewitnesses said the teeth were like a human's, while a fourth simply described ape-men's teeth as "ugly."

The association of sighting reports with a more humanlike creature is more evident from details of the hair. I'm assuming that a hominoid that, in other respects, looks more like a modern human than a monkey would also be less hairy and would have head hair longer and thicker than body hair, especially if the hair was straight (as Lio always described ape-man head hair to be). We can never be certain whether non-sapiens hominins like *Homo floresiensis* were hairy or not. Yet according to one theory, even in the tropics hominins are unlikely to have lost heat-retaining body hair before becoming clever enough to replace this with fire, body coverings, or shelter to keep warm after sunset.[1] There's no evidence that floresiensis had any of these things, nor is it clear that their tiny brains would have facilitated their development. What is certain, though, is that nights on Flores can be cool, even cold, especially in the highlands.

As for Lio ape-men, not a single non-eyewitness mentioned head hair longer than body hair. In fact, all popular accounts suggested short head hair, with most describing this as no longer than the body hair. A minority of non-eyewitnesses spoke, for example, of body hair "like a human's," "sparser or shorter than a monkey's," "shorter and sparser than their short monkeylike head hair," or even "no body hair." All these descriptions, even the last, could be different ways of referring to the same condition. But however interpreted, they obviously contrast with details given by eyewitnesses. Of eleven eyewitnesses who reported on both head and body hair, seven described head hair longer than body hair. Of the remainder, three said head and body hair were the same length, one described the head hair as thicker but no longer than the body hair, and one said the body was hairless while the head hair was short. (Ura, too, described a hairless body and short head hair, but his account had all the hallmarks of a dream.) The color of the hair did not differ greatly between eyewitness and non-eyewitness descriptions, nor did color of the skin.

Eyewitness and non-eyewitness accounts reveal fewer differences concerning other physical features, but this is mostly because these were often not mentioned at all. This observation particularly applies to lower parts of the body. For sighting reports, the absence is understandable when we consider that someone confronted by a strange, erect-standing creature will naturally be drawn to the face and head—particularly when the creature is ambiguously human. A case in point is the feet, mentioned in any detail by no more than three eyewitnesses, all of whom described subjects' feet as humanlike but proportionally long. Only a single non-eyewitness mentioned feet, which he described as monkeylike and used like hands to grasp tree branches.

Of all parts of the lower body, people most often remarked on sexual features. Six eyewitnesses mentioned female breasts, all specifying these as like humans'. Similarly, three non-eyewitnesses said the breasts were humanlike, while one described hominoids as possessing animal-like nipples resembling a dog's (as did one secondhand account). Four of five non-eyewitnesses also said the genitals were humanlike. However, none of the eyewitnesses had noticed their subjects' genitalia, or for some other reason did not mention these. In a few instances people explicitly stated that genitals were not visible. Otherwise their absence from eyewitness reports could be put down to the brevity of encounters, observers being more drawn to the upper half of the body (hence the six references to the breasts), or informant embarrassment concerning sexual organs.

Among eyewitnesses, greater attention to breasts is further explained by the larger number of females that were observed—an outcome obviously contradicting the idea, mentioned in two popular accounts, that only males are ever seen. Yet a closer look at specific encounters reveals a slightly different picture. Where sex was specified, four eyewitnesses reported female subjects while another three specified males. One man (Dapo, chapter 6) identified what he saw as female solely on the basis of a companion's interpretation of the subject having recently given birth. And another two eyewitnesses (Wolo and Noko), after first identifying their subjects as female, later said they weren't sure about the sex. Three other observers reported seeing pairs or groups of hominoids that included both sexes. All the rest, 16—or around half of all eyewitnesses—did not notice or never mentioned the sex. Part of the reason could be that the sexes don't differ much to a human eye. But additional factors likely include circumstances that made it difficult to observe the subjects fully.

Still, eyewitness descriptions of ape-men's sexual features lend further support to their depiction as creatures more like modern humans than like monkeys. So what might we take from all this? For one thing, the differences suggest that most eyewitnesses were not describing monkeys but something decidedly more humanlike. (Recall that people who apparently did see monkeys described these as very much like monkeys.) For another, they suggest that eyewitnesses were not falsely or inaccurately reporting sightings of an unfamiliar hominoid and then drawing on a popular stereotype to fill in the details. Conversely, non-eyewitnesses are more likely to infer specific monkeylike features from general knowledge that ape-men are somehow monkey-like. For example, having heard that ape-men look more like monkeys than do modern humans, some non-eyewitnesses may simply have inferred that particular features—such as head hair, eyes, ears, or teeth—resembled those of local long-tailed macaques.

We should also not forget that most eyewitnesses recounted experiencing fear upon meeting with ape-men, even dead ones—a reaction hardly usual when Lio encounter monkeys. What's more, the most emphatic accounts of negative emotional reactions formed part of sighting reports that were most compelling on other grounds. To be sure, Koli and Watu, the young hunters who undoubtedly encountered a monkey, also reported being afraid. But in this case what unnerved them was not the creature's physical appearance but its apparent invulnerability to their air guns and its appearing to "disappear." Of course, none of this tells us what the humanlike subjects of other sighting reports were. If not monkeys or civets, some or all could have been anatomically modern humans—a possibility I explore later.

Before moving on, we should take another look at secondhand accounts—observers' stories narrated by others. Although far fewer in number and of varying detail, such stories nonetheless suggest several generalizations when compared with both eyewitness reports and popular descriptions. Of six secondhand accounts that mentioned the subject's height, five indicated an erect creature standing a meter or more, including what was apparently the tallest, at possibly 1.45 meters (4 feet, 9 inches). (The tallest eyewitness estimate, we might recall, was 1.03 meters, or 3 feet, 5 inches.) For hairiness, details were less complete. Although five narrators commented on the hair, only one mentioned both head and body hair, both described as sparse. Of the rest, two described a bald head, while another spoke only of sparse body hair. Just two accounts specified the subjects' sex (one concerned a male, the other a male-female pair), but both narrators described sexual features that were humanlike.

Indications are, then, that the subjects of secondhand reports were decidedly more like humans than like monkeys. In fact, especially with reference to height, the accounts might suggest creatures more humanlike than even eyewitness accounts. Seké's hominoid encounter involved a pair of creatures that, standing about a meter tall and with sparse hair, seemed to differ little from ape-men as depicted in both eyewitness and popular accounts. Yet, in this case, too, the narrator said Seké had told him that, when he first saw the pair, he "thought they were human."

Like the original authors of other secondhand stories, Seké further described how the creatures had frightened him. Indeed, their initial appearance caused Seké and his wife to flee. Of course, this and other such accounts also included more questionable details, including hominoids speaking in the Lio language. But we don't know that these were not embroidery supplied by the narrators—as would appear more likely than narrators forgetting or deleting fantastical elements contained in the original story. Anyway, if it were not for such details, the narratives would appear as credible as eyewitness reports.

The Limitations of Observers' Descriptions: Memory and Other Factors

Sighting reports of ape-men have an obvious value. Certainly, if no Lio claimed ever to have seen one, we should be much more skeptical about their existence. Yet eyewitness testimony has obvious limitations. Questions of dishonesty and deception were discussed in chapter 1. As I pointed out there, deception requires effort and motivation. It is much easier to claim ignorance of a topic. And since for the most part what Lio, both eyewitnesses and others, say about ape-men is the same—people typically describe a hominoid that in several respects appears intermediate between humans and monkeys—skeptics wanting to charge deliberate fabrication would need to propose a virtual conspiracy.

A different issue, and one I've yet to address, is memory. Some sightings reputedly took place two to five years before I recorded them. The most recent occurred just a few weeks before. Yet many others were far older, including some of the most compelling encounters, two of which were set in the late 1960s or early 1970s.

In recent years much has been written on the shortcomings of human memory, and it is now clearer than ever that a memory, specifically the memory of a particular event, can fail at any stage—from the moment it is first acquired and stored in the brain (or "constructed") to any subsequent occasion when it is recalled. In the present context, for example, a person might come across a large monkey and, for whatever reason, misinterpret it as an ape-man and carry this memory around for their rest of their life. Or they might initially take the subject for a monkey but later—perhaps after relating their experience to others and hearing their interpretations—decide that it wasn't a monkey after all.

For these reasons eyewitness evidence has sometimes been called the worst sort of evidence, particularly in legal cases. Nevertheless, courts still rely on eyewitness evidence, and when tested against evidence of other kinds or compared to testimony from other witnesses, it often turns out to be correct. Obviously in our daily lives we depend a great deal on memory, both our own and other people's—so much that it would be difficult to imagine what social life would be like if human memory were not reasonably reliable, at least a lot of the time.[2]

All that said, much depends on what sorts of memories, or what components or aspects of a memory, we are talking about. For a start, with visual memories people are more likely to remember the kind (or "category") of thing they saw rather than specific details. Thus we'll recall seeing a dog rather than a cat, or two cars colliding rather than someone being run over, whereas we may misremember or be unable to recall the color or size of the animal or the make of either vehicle or the apparent age of the victim. We may also misremember when we saw something, or what we were doing at the time—parts of the setting or context of the experience. We may even recall seeing something that we only heard about from someone else.

Still, it's fair to say that most of the time people who think they've witnessed something have actually witnessed it. And if they accurately identified the thing in the first place, chances are they will retain a reasonably accurate memory. Of course, memory formation can be affected by poor visibility or mental impairment. We've already seen how some Lio took monkeys for ape-men, so mistakes or misinterpretations are certainly possible right from the start. But how likely is it that all sightings of ape-men we've reviewed involved misidentified or misremembered monkeys or similarly common animals—especially when the subject was at close range or lying dead? We'll come back to this question later. For the moment there are other issues.

Though we don't always recall specifics or background details of visual experiences particularly well, sometimes the context has a positive effect. For example, whereas I came across pigs all the time during my visits to Flores, the one I remember best was a wee piglet that ran inside a house where I was staying. In fact, the animal entered the dwelling several times, and since pigs are not allowed inside Lio houses it was chased out on each occasion. As this illustration should suggest, the unusual character or novelty of an experience is another factor in how likely something is to be accurately remembered. Thus psychologists have found that the "distinctiveness" of an event, either the content or context, facilitates both its retention and retrieval.[3] All the same, while I well remember the size and color of the pig and the exact location of the house, I would have to check my field notes to confirm the year (sometime between 2015 and 2017) in which the episode occurred.

Intrusive piglets aside, another case in point, and one particularly relevant to animals of an unusual kind, concerns the small number of Lio who reported encountering either rare freshwater turtles or equally rare giant coconut crabs, typically on a single occasion. Describing sightings that occurred years or even decades previously—in one instance sixty-five years before—the observers were able to describe specimens in such detail (with regard to form, color, size, behavior, and location of the sighting) that it was possible to match their descriptions with species zoologically documented for other eastern Indonesian islands. And for this reason, my conclusion that these encounters reflected the probable presence of these species on Flores was accepted by peer reviewers and the editors of two international zoological journals.[4] My argument, then, is that what applies to the crabs and turtles applies to equally rare or rarer hominoids. Most Lio have never seen an ape-man, while the large majority of people who've done so claim only a single sighting. So for all but a very few observers, the experience was an unparalleled, once-in-a-lifetime event.

Even so, questions remain about the accuracy of eyewitness memories. On this score, we can do no better than recall those sightings where I was able to speak separately to two people who'd seen ape-men on the same occasion (Bata and Wula, Noko and Dhiki, and perhaps Jata and Ari and Tegu and Wea) and whose descriptions were fundamentally the same. Another case is the two Ngadha men who told me about the extraterritorial sighting discussed in the previous chapter, some twenty-five to forty years after the event. Also, I was able to re-question most of these people one or more years after first speaking to them and, again, what they related on different occasions was not significantly discrepant. What's more, there was no reason to believe that any of these double observers had actively colluded in their testimony. A case where details of what someone told me varied on different occasions involved a single informant—Wolo, the elderly man whose encounter with an ape-man occurred some forty years previously, after the truck in which he was riding struck and killed one. Given the time that had elapsed, and the informant's age and state of health, such inconsistency is hardly surprising. But there's no reason to believe Wolo was not describing the same incident and (notwithstanding the strange turnabout during our final meeting) the same mystery creature.

Apart from the observations of the two Ngadha men, probably the most telling indication of accurate recollection is the testimony of Koli and Watu, who came across what they took to be an ape-man while out hunting monkeys. Koli first told me his story in 2005, just a year or two after the experience, and when he retold it ten years later, the details matched to a surprising degree. In addition, the version given independently by Watu, whom I first met in 2015, hardly differed. As already explained, I've no doubt that what the two young hunters actually saw was, ironically, a large monkey—though one that appeared to behave in peculiar ways and, for this reason, frightened them.

Another variable affecting recall accuracy is whether or how often a witness has told other people about an experience. Previous tellings and retellings involve recasting a visual memory in words, in effect, reinterpreting an event on two or more occasions. Additionally, relating a past experience to others can evoke a verbal response, and this can lead narrators to subsequently revise their accounts.[5] As noted, only when Koli and Watu told others about their disturbing encounter with a monkey was it suggested to them that they'd come across an ape-man—an interpretation they'd evidently accepted by the time Koli first related their experience to me.

The accuracy of some memories can also decline when people regularly experience the same kind of object or event, since they may then confuse or conflate different instances. With few exceptions, however, people come across hominoids only on a single occasion, so this factor is hardly relevant. Because Lio don't always tell people, even close family members, about ape-man sightings, regularly relating such experiences to others also has a limited bearing on their accuracy. Some observers (for example, Malo) told relatives and neighbors about what they'd seen just once, immediately or a short while afterward. Although he may not have been entirely correct, Raga was certain that the story Lalu told me was the same as what Lalu (and his by-then-deceased co-witness, Ndori; see chapter 6) had told him on more than one occasion. And both Wolo and Tegu (chapter 7) claimed they'd never told anyone about their experiences. All the same, some psychological studies indicate that telling others about past events, rather than causing recollections to change, can have the opposite effect. Specifically, such "rehearsal" contributes to the "freezing" or at least slowing down of "the normal process of forgetting."[6]

This brings us to the issue of emotional trauma and its effects on memory. As we've seen, people reporting sightings of ape-men often described not only what they saw but also how the subject frightened them, sometimes severely. The same applies to Lio sightings of rare freshwater turtles and coconut crabs. With these animals, emotional disturbance is, by all indications, caused by suddenly and unexpectedly coming across a strange creature one may be barely familiar with, if familiar at all. There's no reason to believe that an ape-man encounter would be any different. If already versed in ape-man lore, witnesses will have heard about the hominoids' great strength. They might also know about the mystical danger such encounters represent, not least because ape-men—like the rare turtles and crabs—are, in effect, things that people are not supposed to see, or even are not supposed to exist. In the Lio view, all turtles and crabs are creatures that belong in the sea yet, as they further explain, on the few occasions when people come across the terrestrial varieties, the sightings always occur far from the sea and often high in the mountains.

For Lio, ape-men are not "out of place" in this sense. Yet they are ambiguous or anomalous in another way, with contradictory elements inherent in the very bodies of the creatures themselves. A person encountering an ape-man sees a naked, humanlike figure that is not a human being, or at least not human in the same way local people are, but nonetheless bears an uncanny resemblance to a human. In consequence, observers would likely experience profound ambivalence manifested as anxiety or fear. Concerning another kind of not-quite-human thing, an interesting comparison can be found in research by Japanese roboticist Masihoro Mori on androids, or humanoid robots. According to Mori, robots that resemble humans only in a general way don't bother people, nor do androids that appear fully human. But somewhere in between, there is an "uncanny valley" where robots look almost human but not quite, and real humans find them unnerving and revolting.[7]

So what does all this have to do with memories of ape-man encounters? There can be no doubt that severe emotional trauma—caused, for example, by experiences of genocide, active warfare, or sexual assault—can lead to serious psychological disability that, if not treated, can become permanent. This applies particularly when people suffer highly traumatic experiences regularly and over long periods of time. However, ape-man encounters are typically single, isolated events, and even though Lio describe them as frightening, it is anything but clear whether any were "traumatic" to this degree. Rather than resulting in defensive amnesia as was once thought, memories of extremely traumatic experiences can occur involuntarily, are typically "fixed," and can manifest themselves as partial and fragmented recollections that don't always result in well-formed memories or coherent stories.[8] But, again, there's no indication that many, if any, encounters with ape-men were traumatic to this extent, let alone that they caused long-lasting psychological damage.

Drawing on Sigmund Freud's theory of repression, psychologists in the late twentieth century promoted the idea that traumatic experiences, especially of a sexual nature, typically result in "repressed," "dissociated," or "hidden" memories. Yet recent research indicates quite the opposite: a "trauma superiority effect," whereby memories of disturbing events are actually *better* retained than other memories, remaining "highly consistent over time," "significantly more stable," and "more resistant to forgetting."[9] Such memories have also been described as "reasonably accurate and better retained" than memory "for more routine experiences." And where negative emotional reactions are caused by "central or salient information" contained in an experience (for example, what sort of extraordinary animal one saw rather than the color of its skin or hair), it is this, rather than more peripheral information, that "tends to be especially accurate and persistent in memory."[10]

So it seems we should place most credence in ape-man encounters where eyewitnesses did indeed report a negative emotional reaction. And, remarkably enough, these include eyewitness accounts that appear most compelling on other grounds. Of course, a number of Lio said they'd seen an ape-man (lai ho'a) whereas what they described pointed to another animal, usually a monkey. But these cases are sufficiently explained by poor lighting, unusual circumstances or behavior or, in at least one case (Kesu), informant embellishment.

Anyway, I spoke with dozens of eyewitnesses, and the majority described a creature unknown to academic zoology: an erect, bipedal hominoid, lacking a visible tail and considerably larger than a monkey, which looked more like a human. So if all or most of these people actually saw monkeys, we need to ask not only why they remembered seeing something quite different but also why they reported seeing basically the same thing? Or, if observers actually saw a number of different things (other animals, odd-shaped rocks, or even other people)—or perhaps, nothing substantial at all (shadows, odd lighting effects)—why do they consistently describe something that corresponds to none of these things? Put another way, where does the image of the ape-man come from if not from something that looks hominoid but not human?

Could Ape-men Be Familiar Animals or Unfamiliar People?

Skeptics should ideally address the questions above. But even if they did, I suspect many would remain unconvinced that ape-men are completely real. Some might stick to a view of the Lio hominoids as entirely imaginary. Others might conclude that what observers describe could have some basis in zoological reality but reflects a creature far more familiar.

In the second case we have to ask what this more familiar thing could be. Because monkeys are the only nonhuman primates scientifically documented in Flores, the most obvious answer seems to be long-tailed macaques. Yet this suggestion is actually easier to dismiss than some alternatives. Lio describe ape-men as facially resembling monkeys, being hairy like monkeys, having a common ancestor with monkeys, and stealing from fields like monkeys. Also recalling ape-men, Lio sometimes say monkeys possess an intelligence comparable to humans. These similarities may sufficiently explain why people occasionally mistake monkeys for ape-men. Yet the minority of sightings where the subject seems to have been a monkey hardly prove that all sightings were of monkeys. Pora, a man who claimed to have seen ape-men, related how a cultivator once called him to look at a creature he'd caught in a snare. The cultivator thought it might be an ape-man. However, Pora's verdict was that it was only an old and very large male monkey.

Indeed, like Pora, the majority of Lio recognize ape-men as physically and behaviorally different from monkeys in numerous ways. They know, for example, that long-tailed macaques are obligate quadrupeds (meaning they necessarily move on four limbs) and that while they can stand up on their back limbs, they never move solely on these. Just as importantly, though, monkeys are extremely common animals in Lio country, both in the wild and as captive pets. In fact, their numbers, which Lio say are nowadays increasing, are a regular source of complaint among cultivators. So anyone proposing that not just all ape-man sightings but the very image and idea of much larger bipedal, humanlike ape-men derive from people's experience of monkeys needs to answer this question: Why should people usually observe, understand, and talk about monkeys as monkeys and yet, on the odd occasion, represent or experience them in a very different way?

There's one more objection to reducing ape-men to encounters with monkeys. Including some of the more credible eyewitnesses (notably, Malo and Bata and Wula), several people said they'd seen ape-men at night. By contrast, Old World monkeys are diurnal creatures—active only in daylight—and, for the same reason, don't usually enter dark caves. But these observations raise a question.

Two coastal villagers described long-tailed macaques—a single specimen in each instance—not just moving about after dark but entering a house or an outdoor kitchen (a smaller structure built behind a house) looking for food. In both instances, the villagers described this behavior as new and a reflection of monkey populations expanding both in the highlands and on the coast. I've often wondered if the increasing use of electric lights in human settlements is a factor encouraging nocturnal activity among Florenese monkeys. Oddly, little research seems to have been done on this, even in large Southeast Asian cities like Singapore or Kuala Lumpur, where long-tailed macaques also occur. Primatologists—scientists who study monkeys and apes—have told me that, because this species is extremely adaptable, they wouldn't rule out the monkeys, in certain conditions, becoming active after dark.[11] But, even if Flores monkeys are, in fact, becoming partly nocturnal, we can hardly attribute all nocturnal ape-man sightings to these small, long-tailed quadrupeds. For one thing, some took place decades ago, well before electricity was available. For another, none of the settings where ape-men were reported after dark was well-lit, even though some of the subjects were partly illuminated by vehicle headlamps or, in one instance, a flashlight.

In several ways, apes are a better bet than monkeys as a source of the ape-man image. Lying east of Wallace's Line, no apes occur naturally on Flores, nor is there fossil evidence for apes having occurred in the past. Unlike monkeys, introduced to the island some four thousand years ago, there is no evidence for humans ever bringing apes to Flores, which then reverted to the wild. Even so, orangutans and several species of gibbons live on Sumatra and closer to Flores, on Borneo. Gibbons also occur on Java and on the Asian Mainland (see figure 9.1). All, of course, are tailless, can stand on their hind limbs, and can walk short distances on two legs. Most of these apes are also considerably bigger than Flores monkeys—including the one-meter-tall siamang (*Symphalangus syndactylus*), the biggest of the gibbons that lives in southern Sumatra and the Malay Peninsula—and they are proportionately far stronger than humans, just as ape-men are said to be. Besides, according to present-day Lio, orangutans and gibbons *are* ape-men insofar as people familiar with these apes, mostly men who've worked in Borneo, nowadays identify them as varieties of lai ho'a (ape-man). Identifying apes with ape-men, of course, confirms the Lio understanding of local hominoids as a natural species more humanlike than smaller, long-tailed monkeys. Yet we obviously cannot conclude from this identification that ape-men are simply apes.

In recent decades, some Lio have been exposed to foreign apes from the odd specimen brought to Flores. Tipa (see figure 9.2), the man introduced earlier who claimed to have seen an ape-man that "walked badly," told me how, in the mid-1980s, he'd seen another lai ho'a inside a cage at a weekly market in northwestern Lio. (Originating in the mid-twentieth century, these markets are a modern institution on Flores.) The creature had accompanied a man from

Borneo, a peddler who dressed in the traditional costume of a native Bornean. In this the man followed the usual practice of peddlers, mostly men hawking medicines supposedly possessing quite miraculous qualities, who typically bring along unusual animals to attract customers. Pythons and baby crocodiles are among the most common. But what Tipa saw was definitely an original.

Described by Tipa as a "tiny human" that lacked a tail and stood erect to a height of about 45 centimeters (1 foot, 6 inches), the captive ape-man was evidently a gibbon. Some of the time the little ape, while standing on two legs, would hold its hands behind its back "just like a human"—much to the amusement of onlookers. Tipa thought the animal was quite young, which could explain its relatively small size and possibly his odder claim that its arms were "short." (Gibbons have extremely long arms.) If it was young, the creature could alternatively have been a juvenile orangutan, yet Tipa described the hair as grayish or brownish rather than red. Exotic short-tailed or tailless Indonesian monkeys like the pig-tailed macaque (*Macaca nemestrina*) and the crested black macaque (*Macaca nigra*) seem far less likely candidates, in the last case because of the hair color advertised in the name. Another possibility is the rare, virtually tailless, and slightly larger Moor macaque (*Macaca mauru*), a creature endemic to southwest Sulawesi, the large, irregularly shaped island to the north of Flores that is also home to the black macaque. The Moor macaque has a male body length of 64 to 69 centimeters (2 feet, 1 inch to 2 feet, 3 inches) and a tail measuring just 4 to 5 centimeters (1.6 to 2 inches)—the sort of length sometimes estimated for the notional tails of ape-men (chapter 2).[12] This macaque also has a protruding dog-like face, giving it the alternative name "dog-ape" and possibly recalling the few instances where Lio described ape-men as resembling dogs. But, as we've seen, there are easier ways to explain this occasional attribution than the incidental and purely hypothetical occurrence on Flores of an exotic monkey.

Although bringing exotic animals to Flores is now restricted by government regulation, occasionally smuggling a lone ape to the island probably wouldn't be that difficult. As Lio friends remarked, it would have been easier during the twentieth century, when regulations were less strictly enforced. Nevertheless, further inquiries both in Lio country and elsewhere about "visiting" apes produced few results. I heard just one other report that seemed reasonably credible. This concerned another Bornean, reportedly not a peddler but someone on government business, who once visited Flores with a pet ape that wore clothes and spectacles and whose description suggested a young orangutan. So it seems that very few Flores Islanders have ever come across exotic apes on Flores itself. If live specimens are seldom seen, ape body parts are just as rare. The only evidence of these was the photograph of an orangutan skull mentioned in chapter 2, which someone had identified as the skull of an ape-man.

Whatever the incidence of people transporting gibbons or orangutans to Flores, any suggestion that Lio ape-men simply reflect exotic apes must be rejected on several grounds. Apart from the natural absence of apes in these eastern islands, it's unclear how apes brought from Borneo or other parts of western Indonesia and released into the wild would fare in Flores's cooler and drier environment. Also, wild apes would be far more noticeable than rare ape-men. Descending infrequently to the ground, both gibbons and orangutans are highly arboreal, and so would be more visible—and arguably more vulnerable to hunters' spears, arrows, blowpipes, and, nowadays, air guns as well. Gibbons especially are very vocal creatures, but though their loud hooting might recall the nocturnal sounds Lio attribute to ape-men, neither gibbons nor orangutans emit such sounds after dark.

Further countering any suggestion that ape-men might simply be exotic apes, we should also not forget that the Lio concept or image of the hominoids must have existed long before the 1920s, when Father Arndt was compiling his dictionary and recorded the name "lai ho'a." By the same token, Lio would have known ape-men probably centuries before the last decades of the twentieth century, when people began traveling abroad as migrant laborers and when Flores Islanders first had access to newspapers, television, and other modern media—including an Indonesian bank note depicting an orangutan issued in 1992. It might still be asked whether, rather than experience of exotic apes, ape-men could reflect an undiscovered species of ape on Flores that has so far eluded scientists. Simply saying this possibility is ruled out by various sorts of evidence might be countered by asserting that neither have scientists found living non-sapiens hominins like *Homo floresiensis* living on the island. But at least there is fossil evidence for such hominins, and we now know that hominins of some sort have been able to survive on Flores for over a million years. In other words, unlike apes, hominins have lived on either side of Wallace's famous line for a very long time.

But this contrast may suggest another possibility. Perhaps ape-men are modern humans (*Homo sapiens*), though ones that Lio hardly ever come across and so are not fully aware of as a distinct, though equally human, population. In other words, maybe the hominoids are something like a "lost," "hidden," or "uncontacted" tribe. There are estimated to be more than a hundred such groups left in the world.[13] If any survive on Flores, they have—by definition—remained culturally, socially, and territorially separate from Lio, and they could be physically different as well. Living in high mountain forests and subsisting on wild foods, as ape-men are said to do, would meet all these criteria.

For other reasons, too, the suggestion may not be as unlikely as it sounds. In highland regions of the Malay Peninsula (West Malaysia), southern Thailand, and the Philippines, there live small populations of hunter-gatherers that anthropologists call "negritos" (Spanish for "small black people;" see figure 9.1). As the name suggests, these are relatively dark-skinned short-statured folk with an average adult male height of around 1.5 meters (4 feet, 11 inches). On the basis of these features, anthropologists have classified negritos as a variety of "pygmies"—a term usually reserved for physically and culturally similar people in Africa. Researchers have even reported some African pygmy populations as having relatively hairy bodies. Anthropologists have not characterized Southeast Asian negritos in this way. Yet we might note that Indonesians displaying other elements of the negrito physical type—though not forming demographically or culturally distinct populations—tend to be hairier than larger, lighter-skinned people.[14]

Whatever the actual physical similarities, a more intriguing comparison with the Flores ape-men lies in the way taller neighboring peoples, in both Africa and Southeast Asia, view negritos and pygmies. For some describe their smaller, forest-dwelling neighbors as apes or as related to apes. Taller Africans, for example, have classified pygmies not as humans but as "'man apes' like chimpanzees." Similarly, Malays consider the negritos of West Malaysia as descendants of gibbons or, more specifically, siamangs—the largest of the gibbons—and the negritos apparently see themselves in the same way.[15] (Lio, we might recall, regard both monkeys and ape-men as descendants of deviant humans.) I've found no evidence that negritos in the Andaman Islands have ever been represented as descending from nonhuman primates. As mentioned earlier, however, outsiders have sometimes claimed these people possess tails and dog-like faces.

If Africans and Malaysians might appear peculiar in this respect, it should be noted that, for a long time, apes and small-bodied humans were also connected or confused in the imagination of Europeans. Before 1870, when the German biologist Georg Schweinfurth settled the question by "discovering" the pygmy Aka people in central Africa, Europeans believed the pygmies they had read about in the literature of ancient Greece and Rome were imaginary. Then, after the sixteenth century, as Europeans became more familiar with Africa and the Far East, the name became attached to apes. Thus the "pygmy" famously dissected by English anatomist Edward Tyson in 1698 was, in fact, a chimpanzee, though Tyson called it an "Orang-Outang" and in his monograph included a picture of his subject walking erect like a human, although with the aid of a stick. Indirectly connected with "Orang-Outang" (a variant of Malay and Indonesian "orang hutan," literally "forest people," the name Lio nowadays sometimes apply to ape-men), "pygmy" still serves as the scientific species name of the orangutan—*Pongo pygmaeus*.[16] A better illustration of how Westerners for a long time confused beings originally thought mythical with animals (apes) and both with humans (pygmies) would be hard to find.

Like Lio ape-men, African pygmies are known to steal from fields of neighboring cultivators, as chimpanzees also do. Indeed, chimps have been discovered stealing crops at night.[17] Both pygmies and Southeast Asian negritos, moreover, were traditionally nomadic, regularly moving between different parts of their territories to take advantage of seasonal resources. So, although the behavior was mentioned by just two elderly Lio, we might recall descriptions of ape-men as not permanently inhabiting single loca-tions. Larger neighbors have also claimed African pygmies can disappear and remain invisible. Southeast Asian negritos have been credited with other supernatural powers, or even identified as spirits. Kirk Endicott, the foremost anthropological authority on Malaysian negritos, says Malays classify negritos with "spirits and animals" and think that "being akin to them" negritos are not "subject to spirits" in the same way as are Malays. He further observes that, in Malay thought, negritos appear to occupy a boundary between humans and nonhumans.[18] Many physical traits, behaviors, and abilities that larger neigh-bors ascribe to negritos and pygmies are, of course, also things Lio sometimes attribute to local ape-men. So what I'm suggesting is that if negritos or other relatively small-bodied hunter-gatherers did, in fact, live in their vicinity, the Lio people would very likely characterize them in the same way.

That said, the idea that ape-men reflect a largely hidden population of small-bodied modern humans fails on several grounds. First, although twentieth-century anthropologists proposed negritos as the earliest human population on Flores, culturally distinct and territorially separate groups of such people have never been found on the island nor anywhere else in Indonesia—at least not in historical times. In fact, the sole instance of anything like a "lost tribe" in Southeast Asia are the Tasaday people of the Philippines, who came to the world's attention in the 1960s. However, while living in caves and subsisting on wild foods when contacted, these people are physically no different from neigh-boring cultivators.[19] Interestingly enough, the missionary-paleoanthropologist Theodor Verhoeven (1907-1990), whose Flores excavations paved the way for later research leading to the 2003 discovery of *Homo floresiensis*, believed that separate groups of negritos still lived on Flores Island during his own time. But as he was actually referring to no more than a "phenotype"—a combina-tion of physical traits found in individuals all over the island—he was quite mistaken. So too, and for a similar reason, was Teuku Jacob, a leading Indo-nesian paleoanthropologist who later argued that the remains of floresiensis were not a new species but belonged to the ancestors of small-bodied people, or "pygmies" as he called them, who currently live near the discovery site. A

2018 study shows that these people are genetically similar to larger-bodied neighbors and, more importantly, reveal no genetic connection with "deeply divergent hominin groups" like floresiensis. [20]

Further arguing against any connection between the ape-men and hypothetical negritos is the fact that, with an average adult male height of around 1.5 meters, negritos are much taller than the hominoids. In fact, they would not be much shorter than many Flores Islanders, whose average male height has been calculated at just over 1.6 meters (5 feet, 3 inches) and whose average female height appears to be some 10 centimeters (about 4 inches) shorter. [21] Also, whereas Lio consistently describe ape-men's head hair as straight, negritos typically have tight curly, woolly, or frizzy hair—as do many Lio. But the main reason for rejecting the idea that ape-men may simply be unrecognized small-bodied humans is the fact that negritos everywhere maintain relations of exchange, trading forest products for cultivated foods or manufactured items with neighboring larger-bodied agriculturalists. Upon meeting, culturally or physically different groups may trade blows. But eventually they're more likely to exchange goods or services—even if the terms may not always seem particularly fair. (Both negritos and pygmies, for example, have been taken by larger neighbors as slaves.) And occasionally they also exchange members of the opposite sex or, in other words, intermarry.

No such relations exist between Lio people and ape-men. In some respects, the suggestion that the hominoids reflect fully human beings may seem preferable to locating the image in odd encounters with monkeys or apes. Yet an origin in monkeys or even apes is superior to the "negrito hypothesis," for monkeys do, of course, occur on Flores—as does the occasional visiting ape.

If ape-men are not completely imaginary, there remains just one other possibility. Like negritos, the presence of these hominoids on Flores has yet to be scientifically recognized. Yet similar beings have left physical traces on the island in the form of the diminutive remains interpreted as *Homo floresiensis*. How far this entity or a similar, closely related, hominin species may be connected with the Lio ape-men is the topic of the next, and final, chapter.

10

COULD APE-MEN SURVIVE
ON FLORES ISLAND?

All that Lio say about lai ho'a (ape-men) sounds like nothing so much as a non-sapiens hominin—that is, a physically "primitive" member of the genus *Homo* distinct from *Homo sapiens*. On Flores Island there is an obvious candidate: the fossil hominin *Homo floresiensis*, sometimes nicknamed "the hobbit." Referring to the ebu gogo of the Nagé region, Mike Morwood, the Australian leader of the floresiensis discovery team who, in this context, cites my 1998 book *Beneath the Volcano*, says these hominoids "fit Homo floresiensis to a T."[1] Actually, he was not quite correct. If anything, it's the lai ho'a—the Lio ape-men—that fit floresiensis to a T (see figure 10.1).

Lai Ho'a and Homo floresiensis

Before going any further we should recall what I said earlier about the virtual impossibility of Lio descriptions of ape-men reflecting published reconstructions of *Homo floresiensis*. As explained, I heard about the ape-men over a year before the discovery of the fossil hominin was announced, and the possibility of subsequent specialist analyses by paleoanthropologists reaching anyone on Flores, let alone Lio villagers, is vanishingly small. Especially in this light, similarities between floresiensis and the ape-man are striking. The type specimen (or holotype) of floresiensis, dubbed LB1, stood about 106 centimeters tall (3 feet, 6 inches) or possibly a bit shorter. But "she"—the skeleton has been interpreted as female—was apparently the tallest of the thirteen specimens found at the single site named Liang Bua. So the heights of all specimens overlap with ape-man stature, often estimated at just above or below a meter (3 feet, 3 inches).

In other respects, too, body size seems comparable between ape-men and the fossil hominin, or at least not discrepant. The short, backward sloping forehead of floresiensis fits descriptions of ape-man faces as monkeylike. Their resemblance to the gaunt faces of elderly people recalls the pronounced toruses (brow ridges, cheekbones) of the Flores fossil and other primitive hominins (including Australopithecines). Floresiensis also lacked a chin. But Lio accounts of ape-men included few details of the lower face. From skeletal evidence we can, of course, know nothing about the hairiness or skin color of *Homo floresiensis*, so no comparison can be made on these counts. Nor does such evidence indicate whether the fossil hominins had any kind of language, an ability apparently lacking in the ape-men as well.

Just as floresiensis was an "obligate bipedalist," meaning fully adapted to walking on two legs (unlike monkeys and apes, which only occasionally stand or walk bipedally), so people describe ape-men as fully erect and normally bipedal. For the fossil hominin, features confirming bipedalism include features of the pelvis and lower limbs. Interestingly, though, other skeletal features suggest floresiensis was a proficient tree-climber and—like Australopithecines and other older hominins—may have spent a good bit of time in trees, gathering food, sleeping, or hiding and seeking protection.

Features of floresiensis that facilitate climbing include thick arm bones that were short but, nonetheless, long in relation to the relatively short leg bones, "almost identical" to those of chimpanzees, as well as strong forearms; "very curved" hand and foot (including toe) bones; and very long feet. Like the feet of apes, floresiensis's foot lacked a longitudinal arch, meaning that their feet were flat. Also distinctive of *Homo floresiensis* is an "archaic" wrist structure, whose carpal bones reveal "a pattern found in living African apes."[2] Could such ape-like wrists, we may wonder, be reflected in the few Lio descriptions of ape-men's hands turning downward when the creatures stand or move bipedally? Possibly. But it seems more certain that floresiensis' wrists, in combination with long feet, short legs, curved toes, and long arms, would have contributed to their climbing ability—not just in trees but also when ascending steep inclines.

Whereas just one Lio mentioned ape-men sleeping in trees, other descriptions link the hominoids with caves—evidently also a haunt of the fossil hominins. More often, though, people describe ape-men as very strong—as strong as apes, one might say—thus recalling the powerful forearms of floresiensis. Popular accounts (but not eyewitness reports) further describe ape-men's arms as relatively long and their legs as short—thus pointing to additional correspondences. Mentioned less often, long toes and fingers, capable of gripping branches (as one man claimed), could reflect the more general idea that ape-men's limbs resemble the limbs of macaque monkeys. On the other hand, the skeletal evidence for floresiensis suggests the fossil hominin could have put its relatively long and curved toes to the same use.

There's more to say about how the floresiensis "identikit" might match Lio descriptions of ape-men. While the fossil hominin stood erect, the structure of the shoulders, including the shoulder blades, collarbone, and upper arms, was quite different from modern humans. To begin with, the top of the upper arm bone was angled only 110 degrees, which is about the same as in gibbons and macaques. For *Homo sapiens* the figures are from 145 to 165 degrees, giving modern shoulders a much greater rotational capacity. In addition, floresiensis collarbones were short, so this—as well as the low degree of "torsion" (twisting capacity) in the upper arms—caused the shoulder blades to slope and appear "more shrugged forward" than in modern humans (except for people with the condition called "short clavicle syndrome"). And the same features would also have resulted in floresiensis's arms being turned somewhat forward, with the elbows further to the front.[3] Whether these features would have made the hominin look "slightly stooped or hunched over," as four Lio (including one eyewitness) said about ape-men, is unclear. Anyway, most accounts described ape-men as standing fully upright, just as did *Homo floresiensis*. Lending stability to the legs while standing erect, the pelvis of floresiensis was broad and "more flared" than in humans. And this would have given the fossil hominins a "potbellied" appearance,[4] thus recalling sighting reports in chapters 6–8 of the abdomens of contemporary hominoids being distended.

The shoulder structure of floresiensis has further bearing on how the tiny hominin would have moved. Susan Larson, a paleoanthropologist specializing in this area of hominin anatomy, has concluded that these shoulders would have rendered floresiensis incapable of running long distances (or "endurance running"). This is because "the backward and forward movement of the arms along with the rotation of the chest" would not have been sufficient to "counterbalance the movement of the legs."[5] All the same, floresiensis would presumably have been able to run short distances, though not at great speeds. (If so, it may have been fortunate for these hominins that the Komodo dragon, the apex predator on Flores then as now, is incapable of running down prey over long distances, and that adult dragons are too big to climb trees.)

How floresiensis stood, walked, and ran suggest more specific similarities with Lio ape-men. Not just the hominin's broad, bony pelvis—similar to those of Australopithecines—but also the short, stout legs have been interpreted as lending further stability to the species' upright posture. Yet they would have been another factor restricting the ability of floresiensis to run fast. So would the hominin's large feet, described as "extremely" or "extraordinarily" long. Of course, these feet were much smaller and shorter than those of a modern adult human. But relative to the species' diminutive size and short legs, they were long indeed—70 percent of the length of the lower legs, whereas the figure for *Homo sapiens* is around 55 percent. Besides, a lot of the length was in the forefoot, including long toes and metatarsals, and the feet were also relatively wide.[6]

Not surprisingly, the Flores hominin's long flat feet would have required them to walk differently from modern humans. References to "flat" or "thin" feet in a secondhand story (Seké, chapter 5) and in an eyewitness report (Lalu, chapter 6) suggest feet without arches and with low insteps. In *Homo sapiens* flat feet, alternatively called "fallen arches," can also cause duck feet (or "overpronation")—a condition recalling one Lio observer's attribution of outward-turning feet to ape-men (chapter 2).[7] As experts have noted, when walking, the very long feet of *Homo floresiensis* would have caused these hominins to bend their knees to a greater degree than modern humans and, in combination with the short legs, to take relatively long strides. Not only would all this have slowed them down a bit, but the length of the feet, especially, has led specialists to characterize floresiensis as moving bipedally with a "high-stepped gait" or, less respectfully, as "walking like a clown."[8] Circus clowns walk as they do because they wear ridiculously large shoes, which means they must lift each leg high off the ground, thus also bending the leg.

Though not many Lio commented specifically on the size of ape-men's feet, two eyewitnesses described long and wide feet and, in one case, also long toes. Another informant described the feet and toes as longer than those of monkeys. But a more intriguing correspondence with floresiensis's long feet and manner of walking is suggested by what people said about how ape-men moved. According to the story related by his sister, Mbira spontaneously referred to the hominoid he saw as "hopping" from one foot to another (chapter 5). Or, as another man commenting on the same story said, "walking by hopping or jumping." Then there are the eyewitness descriptions of ape-men "walking badly" and moving away (from the observer) bipedally, "lifting their legs in a sort of leaping fashion" while retreating up an incline (Lalu, chapter 6). One man also interpreted a sentence in Arndt's dictionary as describing a lai ho'a (ape-man) as moving "on the toes in a hopping (or jumping) fashion"— a movement reminiscent of the long toes of floresiensis and (though not often mentioned) possibly of ape-men as well (chapters 2 and 6). We might additionally recall extraterritorial sightings of erect bipedal hominoids "bob-bing up and down" and moving with a "jumping or skipping gait."

As for the few descriptions of the Lio hominoids appearing to "glide" as they walk, this feature finds no clear mention in published reports of *Homo floresiensis*. As shown, gliding could reflect media images of imaginary crea-tures sailing through the air like Superman. But if it refers to a compliant gait (see chapter 3)—illustrated by the famous "Groucho walk"—we should recall that this sort of gait involves greater knee-bending, raising the legs, and longer strides than does the "stiff" gait characteristic of modern humans. Whereas such stiff movement allows greater recovery of stored energy on a level surface, a compliant gait should facilitate easier movement in trees and in the steep and uneven terrain of Flores's interior. As suggested earlier, other descriptions of the way ape-men move on two legs, for example "walking as though jumping or hopping," could equally reflect Lio interpretations of a compliant gait. But more remarkably, walking with such a gait—with feet raised high and legs bent—evidently accords with the ambulations of *Homo floresiensis*, as indicated by the fossil hominin's feet and features of the legs and pelvis.

Besides hopping or gliding, one man—the enigmatic Rawi—described ape-men as capable of jumping over greater distances than can humans. Whatever these descriptions exactly refer to, any such form of locomotion cannot be attributed to Flores monkeys, since long-tailed macaques can stand on two legs but do not move bipedally in any manner. Though the literature includes no mention of floresiensis's ability to jump, it's not difficult to see how long feet would facilitate jumping. One study shows that, in modern humans, jump height is positively correlated with the length of the toes—regardless of a person's height, body mass, lower leg length, or overall length of the foot.[9] So if the fossil hominin was anything like *Homo sapiens* in this respect, it's specifically its long toes that are noteworthy. (Heel length is another factor in jump height, but heel bones were not found among floresiensis remains, and Lio never mentioned ape-men's heels.) How far the little hominin's flat feet and short legs could have offset any advantages long toes provided is uncertain. But, again, it's clear how proficiency in jumping, like skill in climbing, could prove advantageous to any hominin in mountainous Flores—when moving on two legs up or down inclines, for example, or jumping between rocks or other raised surfaces while moving over rugged ground.

In terms of technology and diet, there's little to separate interpretations of *Homo floresiensis* from what Lio say about ape-men. Lio do not attribute fire or tools of any sort to their hominoids. Some authorities earlier credited both stone technology and fire to floresiensis. But more recent research has cast doubt on these interpretations. Although stone tools have been found at Liang Bua, the species' discovery site, it is not definite that these were made by the tiny hominin, since the makers could have been modern humans. As archaeologist Peter Bellwood has remarked, Indonesian lithics show considerable continuity throughout the Pleistocene (the geological epoch that ended twelve thousand years ago) and into the Holocene, so in the absence of skeletal remains exclusively associated with the tools, it's not possible to link particular tool types with different hominin species. And similarly, while Mike Morwood thought floresiensis used fire, evidence for this too remains unconfirmed.[10]

Hominins of the same genus as ourselves that lack stone tools and fire—and probably language as well—may be difficult to accept. Referring to possible mental capacities, Morwood says floresiensis probably had language, whereas Dean Falk, the specialist on the hominin's tiny brain, makes no mention of language.[11] Yet, as noted earlier, not all paleoanthropologists have accepted the classification of floresiensis as a species of *Homo*. Of course, without fire the little hominins would have had to consume both plant and animal food raw—like the chimpanzees and Australopithecines they resemble in several physical respects. Such a diet matches the teeth and jaws of floresiensis, which show "tooth wear suggesting forceful mastication" including "mastication of uncooked meat." Yet researchers have also described floresiensis dentition as "consistent with a tough, fibrous diet," and on circumstantial grounds have concluded that the hominins' menu was dominated by plant foods.[12]

What little Lio told me about ape-man teeth and jaws is obviously not sufficient to reach comparable conclusions. Still, people characterized the hominoids as mostly eating plant foods, mentioning stolen chickens and pigs and wild meat (junglefowl, frogs) as less regular components of their diet. Also described in a few instances (including two possible ape-man sightings) were large canines. Direct evidence for prominent canines in *Homo floresiensis* is lacking; in the type specimen, which has the most complete set of teeth, both upper and lower canines are worn level with the incisors. Nevertheless, pronounced canine juga (bony eminences or vertical ridges over the upper canines reflecting a large canine root) could well indicate large canines.[13]

Presumptions of Extinction

How the fossil hominins resemble Lio hominoids, both physically and behaviorally, should be clear. In fact, there is no feature of either that precludes the names "lai ho'a" (ape-man) and "*Homo floresiensis*" referring to the same thing. Yet one common assumption seems to rule out any possibility that Lio apemen are surviving members of the species *Homo floresiensis* or their latter-day descendants. Known only as a fossil, the hominin is regularly asserted to be extinct. Originally, researchers interpreted the species as finally disappearing some twelve thousand years ago. A recent geological restudy of Liang Bua, the western Flores discovery site, has moved the date back to sixty thousand years ago. Stone tools found at the site and possibly attributable to the hominin suggest a more recent date of fifty thousand years ago. But this is still some forty thousand years older than the original date.[14]

There is, however, an important qualification. The discovery site is the only place where remains of floresiensis have been found—on an island with an area of some 14,000 square kilometers (5404 square miles), the tenth largest in Indonesia. Remains of similarly small-bodied hominins have been excavated in central Flores, at a site named Mata Menge, with a much earlier date of seven hundred thousand years ago. Indeed, evidence of stone tool manufacture from the same region points to a hominin presence on the island before one million years ago.[15] Whether these very ancient central Flores hominins were the ancestors of the inhabitants of Liang Bua has yet to be determined. But whatever the connection, the older remains show that diminutive hominins like floresiensis did not live only at a single site—something that would seem impossible anyway. Mata Menge lies about halfway between Lio country and the floresiensis discovery site at Liang Bua. And evidence that such hominins survived for hundreds of thousands of years suggests that they occupied many parts of Flores—and perhaps other Indonesian islands as well.

Particularly in this light, the idea that such a long-lived species or line of hominins met its end fifty or sixty thousand years ago at Liang Bua reflects no more than an unstated paleontological convention. This could be expressed as "a species known only from fossil evidence should be presumed to have survived only to the most recent date, even if that date only belongs to a single site." It is rather as though space aliens, peering at earth through a very advanced telescope and spotting their first motor vehicle (something they would, no doubt, consider a very primitive machine) assumed, in the absence of evidence to the contrary, that this was the last one ever made. The date then stands until a more recent date is found, or until a living specimen turns up—something that on Flores has happened with some fossil rats.[16]

Fortunately, though, some paleontologists take a more realistic view. Dick Fox, a paleontologist colleague at the University of Alberta, has formulated what has become known among his students as "Fox's Law." This reads: "a species originates before its first appearance in the fossil record, and goes extinct after its last appearance!" Paleoanthropologists researching *Homo floresiensis*, too, have shown willingness to contemplate the species' more recent survival. Even the authors announcing the revised date of fifty to sixty thousand years ago state that whether the hominin survived after this time "is an open question." Not just that, Mike Morwood and other members of the discovery team placed some credence in the possibility of floresiensis surviving into historical times.[17] Richard Dawkins apparently envisaged floresiensis being still alive when, shortly after the discovery, he suggested someone should start looking for a living specimen.[18] However, this was when the extinction date was pegged at around twelve thousand years ago and, sadly, Morwood died in 2013, before the redating. What's more, even though some paleoanthropologists may be able to accept the possibility of a far more recent date for the hominins, they apparently still consider the species extinct.

If researchers wanted to find out how long floresiensis survived and overlapped with modern humans, there's always the option of searching for remains at other sites on Flores. But paleontology (and thus paleoanthropology) by definition concerns organisms that lived in the past, especially ones that apparently no longer exist. It is particularly concerned with the origins and earliest development of species and less about their most recent manifestations. By the same token, paleontology does not investigate living organisms, which of course is the mandate of other disciplines. Nor do paleontologists typically go in search of possible present-day or historically recent survivors of presumably extinct species. So it's hardly surprising that, after finding floresiensis at a single site, paleoanthrologists concentrated their efforts in central Flores's So'a Basin, where since the 1950s, numerous stegodon fossils have been found. Their aim was to search for hominin remains that could link floresiensis with hypothetical Mid-Pleistocene hunters of the ancient elephants and thereby shed light on the possible origins of the late Pleistocene hominins found over ten years earlier in Liang Bua cave.

In searching for Mid-Pleistocene hominins, the researchers followed the lead of the Dutch missionary and amateur archaeologist Theodor Verhoeven, who first unearthed what he believed to be stone artifacts in the So'a Basin in the 1960s.[19] Their efforts obviously paid off. Yet just as remarkable was their diligence. Beginning in 2004, at the So'a Basin site of Mata Menge, the quest continued for years without result. Then in 2010, the search intensified. Local people were employed to dig six days a week and bulldozers were brought in to clear an area of 2,000 square meters (21,500 square feet). The team leader—Gert Van den Bergh—had nearly given up hope when a seven-hundred-thousand-year-old hominin molar was found in October 2014. Also at Mata Menge the team eventually excavated a fragment of a mandible and a further five teeth belonging to at least three small-jawed and small-toothed individuals.[20]

The searchers were not only diligent; they were also lucky. For an even more important factor detracting from the apparent certainty of extinction dates is the extraordinary improbability of finding fossil evidence of any species in the first place, and thus the highly unrepresentative character of the fossil record as a whole. It is estimated that over 99.9 percent of organisms are never preserved, or "compost down to nothingness," and that of the 0.1 percent that are not, "the chances of being fossilized are very small." According to another estimate, "only one bone in a billion ever becomes fossilized." Consequently, only "one species in ten thousand has made it into the fossil record," and some 95 percent of these are creatures that lived under water.[21]

Even if a species does join the fossil record, this doesn't mean it stands a good chance of actually being found. So it is obvious that, if floresiensis or any other non-sapiens hominin did survive to the present or recent times, the chances of finding physical remains would be exceedingly slim. Equally evident is the preponderant role good fortune and accident play in constructing the record of all the animals that have ever existed. For all these reasons, determining with any accuracy when any species not known as a living creature became extinct is similarly difficult. And finding remains of the very last living member of a particular species, after deliberately searching in all places and geological strata where the specimen might be found, would be virtually impossible.

Nevertheless, on finding a new fossil species and assuming that the species became extinct around its most recent discovery date, paleontologists will normally look for reasons for its extinction. A usual suspect is ecological change caused by climatic alteration, natural catastrophe (e.g., volcanic activity), loss or depletion of food sources, or competition with other species newly arrived or becoming more numerous within the species' range. With nonhuman animals another possibility is the effect of expanding human populations, a factor also applicable to non-sapiens hominins such as Neanderthals. When the recognized extinction date for floresiensis was twelve thousand years ago, explanations for the species' demise included the extinction of stegodons and erupting volcanoes. The same hypothetical causes might still apply to the earlier dates currently recognized. However, the timing of stegodon extinction remains subject to disagreement and, anyway, it's not clear how far stegodons figured in floresiensis's diet. There were certainly many other animal foods available after this time—including various species of rodents, bats, birds, and reptiles (monitors and smaller lizards, snakes, and turtles), as well as insects, freshwater crustaceans, mollusks, and gastropods—not to mention wild plant foods. Indeed, modern humans on the island continue to exploit many of these to the present.[22]

We should also not forget that, owing to its remarkably small body size, a species like floresiensis would need far less food than a larger hominin. Writing on the remarkable persistence of tiny hominins on Flores, living on the island for probably one million years, Robin Dennell and colleagues have calculated that, regardless of whether diminutive size reflects island dwarfing or a species that was tiny to begin with, the small body of floresiensis would have cut dietary requirements to half that of a modern human.[23]

During the hominin's occupation of Liang Bua Cave, the only large predator on Flores was the carnivorous Komodo dragon, as it is at present. But while there is no evidence of the giant lizards' killing and eating floresiensis—something the dragons could easily have done—it's clear that tiny Flores hominins coexisted with dragons for hundreds of thousands of years, perhaps assisted by the hominins' earlier-mentioned climbing skills. Also, since Komodo dragons have long been restricted to very dry coastal environments, if any mountain-dwelling non-sapiens hominins were alive today, they would rarely encounter the giant monitors. Now inhabiting most coastal regions, modern humans are, of course, a different matter. With dogs, iron spears, and other weapons, modern Florenese are well equipped to deal with dragons and, indeed, are the main factor threatening to bring this species' long survival to an end.

As for volcanoes, major eruptions as well as seismic events occur all the time in this part of the world. Two well-known nineteenth-century eruptions were Mount Tambora (on Sumbawa Island just to the west of Flores) and Krakatoa (more correctly Krakatau, between Sumatra and Java), but neither of these resulted in extinctions of large mammals. *Homo floresiensis* evidently survived the eruption of Mount Toba on Sumatra seventy-five thousand years ago and, anyway, the previous theory that this "supervolcano" caused a "bottleneck" in human evolution is now disputed.[24]

Quite another matter is the arrival of modern humans. In connection with the revised dates for Liang Bua, it has been suggested that *Homo sapiens* was responsible for the extinction of floresiensis. However, the earliest confirmed date for modern humans on Flores remains eleven thousand years ago. Evidence from other large eastern Indonesian islands suggests this date may be too recent. But so far it is the only date available, and it is not impossible that modern humans did indeed settle on Flores relatively late. Whatever the date, Mike Morwood has suggested that the first modern humans may have occupied coastal regions and only later moved inland to areas inhabited by floresiensis. When that "later" date was remains a question. Even Lio highlanders have not moved so far inland as to occupy their highest mountain forests. Anyway, in another place, Morwood and colleagues conclude that interaction between humans and the smaller hominins "may have involved little or no direct contact, symbiosis, competition or predation."[25]

There are other reasons to question the notion that incoming *Homo sapiens* were the main factor in floresiensis's disappearance. Such displacement could result either from human immigrants killing off their smaller cousins—a scenario apparently echoed in ape-man extermination myths discussed in chapter 4—or from competition over territory and resources in which modern humans prevailed. Of course, the two processes would likely go together: modern humans might kill floresiensis to reduce competition, it being far less likely that ancient humans would exterminate the diminutive hominins without practical reasons for doing so. But, as Morwood and others have suggested, it is quite possible that the two species occupied separate territories, as Lio people and ape-men are depicted as doing at present. Similarly, immigrant humans, exaggerating the powers of the archaic hominins, natural or supernatural, and perhaps repelled or unnerved by their ambiguously human appearance, could have sought to avoid them—just as mutually fearful Lio and ape-men are said to do at present. On the other hand, it should be considered that, initially, relations between floresiensis and modern humans were more positive. Evidence from Red Deer Cave in southwestern China suggests that another small, physically primitive hominin, which survived until at least 11,500 years ago, not only "roamed the same landscape as modern humans in China for about 60,000 years" but also interbred.[26]

To what extent immigrating humans would have affected floresiensis' subsistence is difficult to say. Regardless of how long indigenous hominins shared Flores with *Homo sapiens*, during that time, or for much of it, both populations would have been food collectors. Yet there's no reason to believe that wild foods would not have been sufficient to support both species—especially if modern humans were mostly confined to coastal areas and enjoyed exclusive access to marine resources. Though archaeological evidence is lacking, modern humans probably did not begin cultivating on Flores until about 3,500 years ago. Some evidence suggests pigs were introduced earlier, but when people actually started raising them is unclear, and when chickens were introduced is unknown. Still, if floresiensis or another non-sapiens hominin survived until modern humans began raising either crops or animals, they would likely have begun stealing these, particularly when people moved into the interior and began cultivating highland gardens. If so, their thieving could have caused a violent response from cultivators. That said, we should note that dogs, reputedly a particular bane of ape-men, may not have been brought to Flores before five hundred years ago. Interestingly enough, this is about the time maize was introduced to Indonesia, so this crop, the main cultivar stolen by ape-men, would not have been available until very recently.[27]

Nowadays Lio don't say much about people pursuing or killing ape-men for stealing crops or livestock. But such action is portrayed in myths, and it is possible that Lio persecuted hominoids for this reason in the past, when they say ape-men were more numerous. What's more, introduction of cultivars and domestic animals is a double-edged sword. For while these could have occasioned competition and strife, they could simultaneously have added to the thieves' food supply. Also, being aware of the threat posed by people and their weapons, an intelligent hominoid would devise ways of keeping largely hidden, including remaining in or retreating to higher elevations and perhaps adopting a partly nocturnal lifestyle. Lio describe ape-men as active at night, and partial nocturnalism could equally have applied to floresiensis, a hominin, we should recall, whose cranium points to mental abilities comparable to much larger-brained hominins. So, in sum, there is no definite reason why a popula-tion of small-bodied hominins—either floresiensis or ape-men—could not have been ecologically viable for millennia, even while sharing Flores with modern humans, and have continued to be so to the present or very recent times.

Paleoanthropologists remain divided over whether floresiensis developed, by way of island dwarfing, from a larger-bodied species like *Homo erectus* or whether their ancestors were already small before arriving on Flores. Yet it's generally clear how the hominin's survival on the island owes much to its small size. If Lio ape-men were descended from *Homo floresiensis*, dated to fifty or sixty thousand years ago, they would likely have undergone further physical change. This applies with greater force if the seven-hundred-thousand-year-old fossils from Mata Menge, in central Flores, represent the ape-man's ancestor. In fact, we can imagine the ape-man descending either from floresiensis or from the Mata Menge hominins via another line. Either way, indications are that floresiensis had become larger than the hypothetically ancestral popula-tion, which could have been 20 to 30 percent smaller. For example, the lateral corpus of an adult mandible found at Mata Menge was "21–28% lower and narrower than in the two existing *H. floresiensis* mandibles from Liang Bua."[28] By contrast, if the ape-men represent a different line of descent from the Mata Menge hominins, then like these putative ancestors they too could be some-what smaller than late Pleistocene floresiensis. Neither the paleontological data nor accounts of Lio informants are sufficient to make this determination, even though the ape-men sound a lot like floresiensis as known from their discovery site at Liang Bua Cave. But the main point is that, given over a million years of hominin evolution on Flores, several lines, or species, could have arisen from the original ancestors, and at least one of these could have been better able to survive into modern times than floresiensis.

"Living Fossils," Questions of Coexistence,
and a "Late Surviving" Unilinearity

In evolutionary theory there's no reason why a species (e.g., a non-sapiens hominin) that emerged prior to another of the same genus (e.g., *Homo sapiens*) should not survive well into the period occupied by the newer species, and therefore that both should not survive to the present. Animal species previously unknown to academic science are turning up all the time. It also happens that species deemed long extinct turn up alive.

One example from Southeast Asia is the Indonesian coelacanth (*Latimeria menadoensis*), a large fish belonging to an entire zoological order (the Coelocanthiformes) that was previously thought to have died out before the dinosaurs. Reporting on a 2021 study showing coelacanths to be long-lived and slow-growing, and to have a record-breaking gestation period of 5 years, a science journalist remarks that the fish "has just about the most extinction-prone profile it is possible to imagine."[29] Yet a specimen of the Indonesian species was found alive in 1997 on the island of Sulawesi, just to the north of Flores. Discovered only a year before, in 1996, another example is the Laotian rock rat (*Laonastes aenigmamus*), a squirrel-sized rodent never discovered in the fossil record and representing a brand-new genus of the Diatomyidae, a family that was supposed to have gone extinct eleven million years ago.[30] Coincidentally enough, specimens of both animals were first encountered fortuitously, for sale in local food markets. More importantly, the finders were Western scientists—a Canadian marine biologist and a British conservation zoologist—who just happened to be visiting the markets. In fact, the marine biologist was, at the time, honeymooning with his wife, who was the first to spot the extraordinary fish. So neither discovery was made in the context of actual zoological research. This last qualification is important for what I say later about the chances of local people reporting scientifically undiscovered species. Both the fish and the rat were already well known to local people, and both had local names. Hardly a rare animal, the rock rat is, in fact, currently listed by the International Union for Conservation of Nature (IUCN) as a species of "least concern." This is the same status accorded to black and brown rats, Flores monkeys (long-tailed macaques), and *Homo sapiens*. So perhaps it wasn't unreasonable for someone to post a 2005 internet comment on the Laotian rodent's discovery amusingly titled "How many scientists does it take to see a rock rat?"

In connection with scientific views on *Homo floresiensis*, the coelacanth and the rock rat have another significance. For both have been declared "living fossils," "Lazarus species," and creatures "out of time." Anyone who thinks these phrases are just popular media soundbites should be alerted to their use in scientific venues, as in "Laotian rodent proves living fossil," a 1998 article in *Scientific American*. The phrases are undoubtedly eye-catching and appealing. But in regard to real possibilities in the natural world, they are also misleading. And though all three are obviously metaphorical, people have a well-known tendency to take metaphors literally and lose sight of their figurative character.

Sometimes applied to the Komodo dragon as well, "living fossil" as a reference to coelacanths and rock rats is an obvious contradiction in terms; the creatures no longer exist only in fossil form but are as much alive as any currently surviving species. The expression also disguises a scientific shortcoming. For rather than focusing on problems in declaring species extinct, it tends to suggest that—like vampires (the "living dead")—the creatures *should have remained extinct*. In other words, it puts the onus on the rat and the fish. "Lazarus species" is more artful. It alludes, of course, to the biblical character who, after being dead for four days, was miraculously brought back to life. By contrast, Laotian rock rats and coelacanths (both the Indonesian coelacanth and another species found earlier, in South African waters, in 1938) have been alive all the time and, unlike Lazarus, have never been "dead" (extinct). So there was nothing miraculous about their recent discovery as living species. And, on a more mundane note, phrases like "Lazarus species" and "living fossils" make coelacanths and rock rats look more like exceptions to a rule than they actually are.

Because *Homo floresiensis* is only known as a fossil species, "living fossil" and "Lazarus species" can hardly apply to the hominin. In the usual sense of "at the wrong time or period," however, paleoanthropologists have treated floresiensis as a being out of time—a view that resonates in the species becoming the namesake of fictional if not mythical "hobbits." Time travel aside, a species being out of time is, strictly speaking, impossible, and it may be partly for this reason that anyone who thinks "extinct" creatures still survive runs the risk of being accused of entertaining an impossibility. Discussing floresiensis, archaeologist Peter Bellwood states that the initial date for the species "raised problems," partly because it meant the hominins would have overlapped considerably with modern humans in Indonesia. He then adds, cynically, that to accept such a recent date, "one would have to suggest either that early modern humans in Flores were very nice and polite to their archaic hominin neighbors for upwards of 40,000 years, or that the dates are far too young."[31]

Obviously, Bellwood considered the first alternative unlikely or impossible. But he was hardly alone, for it was evidently the same view that motivated the later geological reanalysis of the discovery site. Thus in their paper announcing the revised dates, the researchers begin by stating that: "A major reason [floresiensis] is controversial is because the *H. floresiensis*-bearing deposits, which include associated stone artifacts and remains of other extinct endemic fauna, were dated to between about 95 and 12 thousand calendar years (kyr) ago." They add that "these ages suggested that *H. floresiensis* survived until long after modern humans reached Australia by ~50 kyr ago."[32] I don't mean to suggest that the newer dates (fifty to sixty thousand years ago) are mistaken or contrived. But we should recall that the dating concerns only a single site (the discovery site) and, moreover, that extinction dates reflect no more than a long-standing convention.

Common paleoanthropological terms like "primitive" and "archaic" also feed into a view of *Homo floresiensis* as being out of time. So too does another common usage, "late surviving." Scientists use "archaic" and "primitive" (which means "first") to refer to physical traits of species that appeared first, or earlier in an evolutionary sequence, and when used in this way, they obviously serve a purpose. But when applied to whole species, the words can introduce a prejudice. For the adjectives weigh against, even tend to exclude, the possibility that floresiensis could have lived longer than the already recent dates to which they have been assigned—let alone living into historical times. As for "late surviving," a phrase that might suggest something is beyond its "best before" date, we must ask "later than what?" Presumably later than the emergence of other hominin species, less primitive or archaic.

Especially when the most recent date was set at around twelve thousand years ago, the discovery of *Homo floresiensis* became Exhibit A for a newer view of hominin evolution. During most of the twentieth century, anthropologists had assumed that more "advanced" species should replace more "primitive" or "archaic" ones in a single line, or unilinear sequence. Toward the end of the century, however, research increasingly revealed hominin evolution as involving several "branches" developing and continuing at the same time, so that several hominin species were contemporaries and the hominin family tree came to appear far more "bushy." In light of this development it is surely ironic that interpretations of floresiensis being rendered extinct by modern humans (or "replaced," to use a nicer term)—whether sixty or twelve thousand years ago—in effect restores the older view of hominin species replacing one another in a unilinear progression, as does the claim that the original date for floresiensis's supposed extinction was "far too young" (or recent).

Equally apparent in suggestions that modern humans must have replaced floresiensis, and closely linked with a persistently unilinear perspective on biological evolution, is another older and similarly contested view, labeled the "competitive exclusion principle." This holds that where two closely related species, especially ones belonging to the same genus, occupy the same environment (or "ecological niche") and compete over the same resources, one will eventually replace the other entirely. Since it is only modern humans who study other hominin species, one might detect a whiff of triumphalism or exceptionalism in competitive exclusion—a position that portrays *Homo sapiens* and their ancestors as outcompeting other hominins all the way down the line. However that may be, the fact is that two species of the same genus often do occupy the same environment, and without any sign of either imminently disappearing. For example, in the Canadian province of Alberta, where I'm writing this, there are two species of bears (genus *Ursus*) and two species of deer (*Odocoileus*). Also, *Homo neanderthalensis*, *Homo sapiens*, and the Denisovan hominins shared Eurasian environments for millennia and even interbred. Evidently, though, the principle does not hold where one species' ecological niche does not entirely coincide with the other, that is, where one species exploits partly different foods or feeds in different parts of a territory, including at different elevations. This, in effect, is how Mike Morwood saw *Homo floresiensis* and modern humans coexisting. It is also what Lio statements suggest for ape-men and themselves. But even where such differences are not apparent, critics of competitive exclusion have produced sufficient contrary evidence for some to conclude that "coexistence rather than exclusion of closely related species is the rule."[33] In any case, it's quite clear how "competitive exclusion," like such concepts as "archaic" and "late-surviving," is an artifact of an older unilinear view of hominin evolution that doesn't sit easily with the newer multilinear, or bushy, model to which floresiensis has provided so much support.

As the exclusion principle concerns the most closely related species, it again raises the question of how closely related floresiensis were to modern humans. In other words, if the fossil hominin is correctly assigned to a genus other than *Homo*, as Peter Brown advocated when he proposed the new genus *Sundanthropus* (chapter 1), then the principle would be less applicable anyway. But whichever taxon was selected, it's obvious that *Homo floresiensis*, with skeletal features reminiscent of hominins not categorized as *Homo* or even of modern apes, and probably lacking stone tools, fire, and language, remains a very different creature from *Homo sapiens*. In fact, in these respects, floresiensis may appear more comparable to equally bipedal Australopithecines, except that some Australopithecines more certainly used stone tools. Mike Morwood, the principal proponent of *Homo*, discusses why the discovery team finally decided on this genus in his 2007 book. This is not the place to assess that decision. However, in view of the lingering influence of unilinearity in evolutionary thought, we can hardly ignore Morwood's statement that the very recent dates for floresiensis were "one more good reason to consider [the species] as Homo."[34]

Why Surviving Species May Not Be Found

Sometimes a different argument is used to counter suggestions that presumably extinct species, like floresiensis, are still alive. Essentially this is the claim that, if they continued to exist, a specimen would have been found by now. The idea goes back at least as far as the French naturalist Georges Cuvier and his assertion, published in 1812, that "there is little hope of discovering new species of large quadrupeds."[35] Because large animals should have more difficulty staying hidden than smaller ones, the statement seems reasonable. But since Cuvier made it, numerous large animals have indeed been found, including, among primates, gorillas. So perhaps all we can be sure about is that undiscovered creatures remain undiscovered only until they are discovered. In the case of ape-men, though, there is rather more to say.

Lio, of course, have "found" ape-men, including specimens people have observed dead or alive. So in this case "discovered by now" must mean either discovered by professional researchers or found by someone else who, directly or indirectly, has reported the discovery to scientific authorities. So why hasn't this happened with the Lio ape-man?

The main reason field zoologists (academic biologists who study animals in their natural environments) have never found an ape-man on Flores is, quite simply, because none has ever looked for one. Whether, as a result of this book, any scientist chooses to take up the task remains to be seen. But there are several reasons such research would be difficult. For one thing, Flores's rugged terrain would require researchers working in the field to be physically fit and relatively young—two factors that precluded this author from going in search of flesh-and-blood ape-men when I began my ethnographic research in Lio, in 2003, at the age of fifty-six. (An experienced senior zoologist might lead a research team, assigning actual field research to physically more able younger colleagues, but these younger researchers would typically be doctoral students or others whose careers are at an early stage and so could be harmed by association with such research.)

Just as importantly, to conduct field research into mystery animals, academics would have to secure funding in the form of research grants, and such grants require rationalizing proposed research in relation to existing knowledge and theory and demonstrating not only the researcher's qualifications (I'm not a trained zoologist) but also a high probability of success in attaining scientific objectives. As regards ape-men, though, there is no existing body of knowledge, at least not in the branches of natural science where the topic should belong. More specifically, there is no recognized theory that would allow for the present existence of undiscovered non-sapiens hominins, nor of new or known apes (orangutans or gibbons, for example) in places like Flores.

In fact, funding problems aside, zoologists—especially mammalogists— typically don't set out to find specific kinds of previously undiscovered animals anyway. They might, of course, encounter new species fortuitously. And in this they have something in common with local people like the Lio, who also come across rare animals quite by chance. But locals have a lifelong familiarity with territories such creatures inhabit; they also have generations of folk zoological knowledge behind them. Field trips by academic zoologists, in contrast, usually last only weeks or a few months.

As regards searching for hypothetical species, what applies to academic zoology applies equally to paleoanthropology. The search for possible ancestors of *Homo floresiensis* may seem to be one of a few exceptions. But this was initially inspired by Verhoeven's discovery of possible stone artifacts suggesting the early presence of non-sapiens hominins on Flores, and the search was conducted after the (accidental) discovery of floresiensis in another part of the island.

Even in the absence of any relevant theory, scientists nowadays would have a better chance of winning support for research into a large undocumented primate if they could come up with prima facie evidence for its existence. But gaining support for a search for living ape-men on the basis of ethnographic evidence alone—that is, what local people, or nonscientists of any kind, say about the creatures—is far more unlikely. To permit scientific documentation, academic zoologists need a body, parts of a body, or a "voucher" as biologists call proof of a species' existence, and this can come from anywhere and be obtained in any circumstance. Good examples are the Western scientists who came across the Indonesian coelacanth and Laotian rock rat in local Southeast Asian food markets. Although Asian consumption of wild animals has recently been blamed for zoonotic diseases like COVID-19, in the case of the rock rat if not the fish, biologists can be grateful for what Westerners would consider "exotic" dietary practices. But the more relevant point is that, in both instances, it was academic biologists who directly came across the new species and who immediately took photos and made efforts to preserve specimens. Local people had been familiar with the creatures for years and presumably generations, yet for obvious reasons they were not motivated to do either.

So what are the chances of local people who come across new species passing this information on to academic scientists? Or since they are unlikely to know what might or might not be a "new species," or to have any understanding of the concept, I should say "local people who encounter animals that to them seem strange or rare." The chances are very slim.

Let's consider a scenario where a Lio person came across an ape-man, dead or alive. For this to effectively reach an academic audience, either the creature would need to be captured and kept alive or, if already dead, the corpse would need to be preserved. Barring that, a good photograph would need to be produced. But none of this is likely. For a start, the person involved (and it is usually lone individuals that see ape-men) would have to be motivated to take any of these measures. It hardly needs mentioning that any Lio who obtained physical or photographic evidence of ape-men would in all probability not realize the scientific significance of what they had seen nor, for this reason, any need to report it to anyone. Preservation of already dead or freshly killed specimens would, to say the least, be extremely difficult in a tropical climate and without necessary chemicals and equipment.

A photo might seem a better bet, but getting a good shot would be difficult in settings where people might come across hominoids. Also, though not many Flores Islanders are yet familiar with the technology, nowadays a picture could be photoshopped and so later dismissed as invalid. Possibly because of his perception of the dead ape-man he found in 2010 and his desire to dispose of it quickly, even Tegu (chapter 7) did not take a photograph. In fact, during all my time on Flores, I've never heard of anyone photographing an ape-man, even though I've heard about photos taken of other unusual things—ghosts, for example. (When I followed up such rumors, by the way, the photos turned out to be either lost or unimpressive—or in the most recent case I encountered, faked.) As some Lio now own cell phones with cameras, this new technology might provide a better prospect for documentation. But I am not optimistic; cell phone photos I saw on Flores were often of poor quality and, after showing them around, phone owners tended to misplace or delete them, regardless of subject matter.

Capturing a live ape-man would not only be difficult but would almost certainly be incidental to an unplanned encounter. The odd claim about capturing hominoids notwithstanding, Lio do not deliberately go in search of ape-men. Besides lack of cause, part of the reason is fear of the creatures, both their natural and supernatural powers—a factor that also applies to keeping a dead specimen, and probably photographing one as well. What applies to capture also applies to deliberately killing an ape-man. As we saw in chapter 4, Lio tell stories from times past of ape-men occasionally getting caught in traps. But whatever one makes of these, stories of accidental capture are rarely heard at present.

Even if a local person did obtain solid evidence of an ape-man—a complete skeleton, say—there's the further question of how scientists would hear about this. We already know that not everyone tells others about their experiences, including two of the most compelling eyewitnesses who described encountering dead specimens. Apart from myself, in recent years the Lio region has been visited by at least two cultural anthropologists, yet the two I know, though familiar with the category "lai ho'a" (ape-man), have never been particularly interested in the creature to which it refers. During the same period, anthropological visitors have included teams of paleoanthropologists involved in excavations relating to *Homo floresiensis*. But their work has been in regions well to the west of Lio country.

As remarked in chapter 7, given the humanlike appearance of ape-men, anyone who found themselves involved in the death or killing of one might fear not just mystical repercussions but also legal consequences and so would be reluctant to report it to outsiders of any sort. If information about any kind of ape-man encounter were to be communicated to anyone besides relatives and neighbors of eyewitnesses, this would most likely be local government officials. I should say straightaway that I've never heard of anyone doing this. We should also recall that even Wolo and Tegu, who observed dead specimens, did not report what they saw to local authorities. But, if any authorities were contacted, these would most likely be local administrative chiefs or "headmen" (Indonesian "kepala desa," or "lurah"), who would quite likely be the reporters' neighbors or kin.

Such headmen, however, would probably take the matter no further. If a sighting involved a dead hominoid (and presuming they found the report credible and worth following up), they would most likely react in the same way as did Tegu and simply recommend disposal of the remains. As I know from over forty years acquaintance with local government officials at all levels, there is far less chance of the report reaching higher authorities—either subdistrict or district leaders ("camat" or "bupati"), either from a headman or directly from the person involved. Providing they were Lio people—and they aren't always—such officials would probably not dismiss ape-men as nonexistent. Yet having received more Western-style education than other people, they would likely be inclined to regard the hominoids as primarily supernatural beings anchored in local tradition. The same applies to Catholic priests, nowadays nearly all Indonesians. And if either did take such a report seriously, it is still not certain that it would later reach scientific authorities.

Local police are another matter. Unless they mistook it for a human being, perhaps, people would be less likely to report a dead ape-man to the police than to government officials or priests, and simply reporting an ape-man sighting would be out of the question. For one thing, by Indonesian government policy, police officers on Flores are usually non-Florenese (often Balinese) and, in my experience, have little understanding of things they consider part of local culture. More importantly, the report would necessarily be lodged in the Indonesian national language, whereby nowadays the ape-man would undoubtedly be described as a "kurcaci," a term understood throughout Indonesia as referring to a supernatural elf or sprite. Actually, the same considerations could apply to district or subdistrict officials, who might not understand "kurcaci" as a reference to any sort of natural being.

We should also not forget that Lio value ape-man bodies as a source of magical relics (Chapters 3 and 4). So, rather than informing officials, anyone coming across a dead ape-man might bury it quietly and later retrieve the bones, either for their own use or for sale. Potential purchasers could include not only local people and non-Florenese outsiders but local administrators. Indeed, I recorded several stories of people running in district elections obtaining ape-man relics to mystically secure a favorable outcome, reputedly for sizable sums.

If a report of humanlike apes or ape-like humans did somehow reach the attention of academics, either Indonesian or foreign, one question is whether they would take this any more seriously than government officials. Clearly much would depend on how the thing was described. I know of just one case where a local report of something like an ape-man was apparently deemed credible by someone with academic credentials. This did not concern Flores but rather the island of Sumbawa, mentioned previously with reference to the supposedly tailed Wawo people and the odd-looking upright primates observed by a Western anthropologist (chapter 8). In 1996, reports appeared in the Indonesian press about a villager in western Sumbawa (the Wawo region is in the east) who had reputedly captured two ape-like creatures. The creatures subsequently died, but the villager kept their skeletons. Somehow his story reached the Indonesian director of the provincial museum of West Nusa Tenggara, the province that includes Sumbawa. According to the same report, the director was planning to visit the villager and inspect the remains.

Whether he ever did so, however, was not subsequently reported, and my later attempts to contact him were unsuccessful. (If the director did pay a visit, the skeletons presumably turned out to be a rumor or the remains of creatures far more mundane, in which event both he and the press would have lost interest.)

Another argument that might be leveled against the existence of ape-men is a variant of the claim that, if they did exist, one would have been found by now. This is the proposition that mystery hominoids, apparently similar to the Lio hominoid, are supposed to exist in other parts of the world and yet none has ever been scientifically documented. I'm referring of course to figures like the North American Bigfoot and the Himalayan Yeti. But there are important differences between these and the Lio ape-man. For one thing, Westerners, including some with scientific credentials, have in fact searched for these creatures. For another, in places where Bigfoot and the like are believed to exist, there are no documented fossil species to which the putative hominoids could correspond—unless we count the exclusively Asian giant ape *Gigantopithecus*, dated to the early to middle Pleistocene in southern China. On Flores, by contrast, we have the remarkably similar species *Homo floresiensis*, which although provisionally deemed extinct, is an extremely good candidate for either an ancestor or a close relative of the living hominoid described by Lio. People might come up with the idea of a being physically halfway between a human and a nonhuman animal on the basis of something other than experience of a natural species. Yet here I'm concerned only with a particular case. If one could show that any mystery hominoid anywhere was a real animal, this wouldn't prove that any of the others was equally real. But neither would evidence casting doubt on the existence of one or more prove that any of the others was necessarily pure fantasy.

Could Ape-men Be Imaginary After All?

So far I have reviewed evidence suggesting that Lio ape-men could be biologically real, all the while taking account of counterevidence and counterarguments. My point of departure has been the Lio view that the hominoids are natural beings or at any rate animals, like dugongs for example, which have bodies and are ordinarily visible, even though they are also credited with supernatural powers. On the same grounds, Lio distinguish ape-men from purely supernatural beings or, in other words, spirits. For the moment, though, let's look again at the possibility that ape-men may be purely imaginary. In that case, of course, the question arises why Lio not only describe them as real creatures but, in some instances, claim to have seen dead or living specimens. As already shown, the argument that people are consistently fabricating will not do. Nor will any suggestion that whole human populations are regularly given to hallucinations.

In anthropology the persistence of ideas that are not backed by any physical evidence (for example, invisible spiritual beings that can walk through solid objects or cause harm by purely psychic means, or the notion that very old people grow tails) has long been attributed to hypothetical symbolic values or social functions. In other words, such ideas are deemed to be social or cultural constructs. Apparently adopting a more psychological approach, science writer Scott Weidensaul, in a book concerning a variety of cryptid animals and people who are convinced of their existence, comes to the rather unoriginal conclusion that people think cryptids exist because they want them to exist.[36] Thus Weidensaul's claim provides a good illustration of a "functionalist" approach to mystery creatures—by proposing that they are invented to serve some (not always specified) human purpose or need.

For obvious reasons I'd be pleased if it could be proven beyond doubt that Lio ape-men were living non-sapiens members of the genus *Homo*. For one thing it would prove me right! On the other hand, the discovery would present us with a tremendous moral conundrum: what to do with something so close to *Homo sapiens* but not quite there? More to the point, it's anything but clear why the Lio would want such frightening, thieving, and threatening creatures to exist. From the mythical notion that ape-men derive from incestuous or otherwise misbehaving humans, we might infer that the Lio hominoids, living an undesirable existence in wild places, symbolize an uncultured state and thus serve as a warning of what might happen if present-day people behaved similarly. Yet not only are such sociological interpretations purely speculative, they gain credibility only to the extent that other explanations, especially the possibility of their reflecting a scientifically unknown species, are presumed impossible. In addition, there is no reason familiar animals, for example monkeys, should not equally serve the same social or cultural ends, especially as Lio credit monkeys with exactly the same mythical origin as ape-men—as transformations of misbehaving humans.

In contrast, anthropologists who take a cognitive approach argue that people are naturally inclined to subscribe to images and ideas that are arresting and thus attractive, and are therefore likely to be passed on and survive, regardless of any social function they might perform. Particularly relevant for this approach are ideas that are "counterintuitive," combining elements that do not normally go together. Yet, again, extant creatures can equally meet these criteria. In fact, according to a highly durable anthropological theory, some familiar animals are symbolically prominent precisely because, in respect to anomalous physical features or behaviors, they strike humans as exceptional and supernaturally powerful.[37] Combining features of humans and animals, it is already clear how ape-men can be perceived as anomalous. So this approach too has no bearing on the zoological reality of these creatures. On the other hand, because people often treat anomalous things as "taboo," the ape-man's anomaly could shed light on the attitude of Westerners who become uncomfortable, even upset, at any suggestion that mystery hominoids or other undocumented animals might actually exist and who apparently wish to banish the very question from rational debate.[38]

Sometimes called "ontological relativism" or "ontological pluralism," a newer approach in cultural anthropology is fundamentally irrelevant to the questions I raise here. ("Ontology" refers to a philosophy concerning what sorts of things exist in the world and how they relate to one another.) Briefly, proponents argue that members of small-scale nonliterate societies like the Lio typically experience—indeed "see"—the world in a radically different way from Westerners and from academic scientists. As far as I know, no ontological relativist has ever addressed the question of how we might understand what people like the Lio say about mystery hominoids. Nevertheless, proponents would be virtually obliged to claim that hypothetically treating ape-men as possibly natural creatures, like scientifically documented animals (as I have done), is completely wide of the mark, for the interpretation has meaning only in the context of what they call "naturalism," conceived as an exclusively Western ontology. Logically, of course, they'd have to treat all other animals Flores Islanders recognize as real creatures—from monkeys to myna-birds to millipedes—in exactly the same way. In consequence, academic science could have no bearing at all on Lio folk zoology and, for the same reason, what Lio know about animals can make no contribution to academic science—a sorry conclusion if ever there was one.[39]

Explanations of Lio propositions about ape-men based on an assumption that the creatures do not exist, at least not in the sense employed by "naturalists," thus leave much to be desired. So we are left with the much simpler explanation that people sometimes see the hominoids because they are actually there—a natural species that has yet to be scientifically documented. A compromise solution might be that ape-men are extinct but survived until recently. If "recently" means until three decades ago, the cause of their demise could have been the 1992 tsunami and a subsequent series of earthquakes, which Lio describe as having resulted in landslides, streams drying up or changing course, and the local disappearance or reduction of more familiar animal populations (notably certain birds). Sightings people continue to report might then be interpreted as what has been called a "shadow," a cultural memory of an animal (perhaps not fully conscious) giving rise to misinterpretations of encounters with other animals—possibly including odd-looking specimens of Homo sapiens.[40] But this is difficult to square with detailed sighting reports of bodies and living specimens recorded almost to the present.

So, after considering the alternatives, a living non-sapiens hominin undocumented by science—something like a latter-day floresiensis or another, very similar and presumably related, hominin—appears to be the best explanation for the Lio ape-man. I must confess that this conclusion leaves me uncomfortable, and I have difficulty fully accepting it. The reasons, however, do not concern the evidence so much as ideological and therefore social factors. Among these is probably a lingering influence of the now increasingly challenged but still persisting unilinear view of hominin evolution, bound up with the nineteenth-century notion of "progress," whose most famous proponent was Darwin,[41] and even older ideas about human perfectibility. Another factor is the stigma deriving from a widespread view of people who entertain the possible existence of creatures not (or not yet) accepted by science as mentally peculiar if not downright delusional! As mentioned in chapter 1, unless treated as purely imaginary, mystery animals like the Lio ape-men are a subject without a recognized academic discipline, so writing about them can be not only a difficult but a lonely venture. But one has to follow the evidence as far as it may lead.

Despite its flaws, it's also hard to resist the view that if mystery hominoids exist, not only would one have been found by now but scientists would have heard about and accepted it. Yet as examples from zoology and paleontology have shown, both living and extinct species can remain hidden from science for a very long time, and there's no reason they cannot remain hidden forever. Besides, there is no statute of limitations on scientific discoveries. The proposition that undiscovered animals do not exist, or will be found only as extinct fossils, is therefore a risky one, for it takes just a single specimen to prove it wrong. By the same token, if only a single one of the Lio sighting reports summarized in chapters 6 to 8 is substantially correct—that is, neither fabricated nor a mistaken identification of a known animal—then the Lio ape-man is an extant species undocumented by scientists and most likely a hominin.

Discoveries of new species also, of course, tend to be unexpected. Though floresiensis is only known to scientists as a fossil, the discovery has been described as "startling" (as in the subtitle of Morwood's 2007 book). Paleoanthropologist Peter Brown took this one step further when he said he would have been less surprised by the discovery of a space alien.[42] Whether the discovery of a living ape-man would be any less startling or surprising than the arrival of extraterrestrials, I leave for readers to imagine.

ACKNOWLEDGMENTS

The idea for this book goes back a number of years. The subject forms part of a longer-term project focused largely on spiritual and more mystical figures of the sort I discuss in chapter 3. What the Lio people call *lai ho'a*, or "ape-men," however, was a topic that attracted me from the time I first heard about them in 2003, and all the more so after the discovery of the fossil species *Homo floresiensis* (aka "the hobbit"). Both in quantity and quality, what local people told me about their knowledge and experience of local ape-men soon made clear that a book length treatment devoted specifically to these beings was both feasible and likely to be of interest not just to scholars and scientists but equally to a wider and more general audience.

My first vote of thanks, therefore, is owed to the many Lio men and women who so generously gave of their time and hospitality to tell me what they knew about ape-men, and indeed other animals, both familiar and rare, with whom they share their native territory. They are far too many to mention individually, and in any case I've seen fit to disguise individuals with pseudonyms in order to maintain their privacy. Nevertheless, special thanks are owed to the family of Venansius Bhara, my main hosts in Lio, who generously provided me with bed and board during a series of visits between 2014 and 2018. The investigations on which this book is based would have been impossible without the support of Indonesian sponsors, in particular Father Philipus Tule, former rector of the major Catholic seminary at Ledalero (Maumere), himself a Flores Islander and a trained anthropologist. On a more practical level, special thanks are owed to Heribertus ("Harry") Ajo, a great friend from the Nagé region of central Flores whose family I have known for many years. Harry has served as my fixer during many visits to Flores, and could always be relied upon to quickly and efficiently arrange air tickets, ground transport, or accommodation when needed—often at short notice.

Closer to home, this book has benefited from encouragement, information, and advice provided by family, friends, and academic colleagues. Not least of these is my wife, Christine Forth, herself a published author and who, like me, holds a doctorate in anthropology. Somewhat jokingly I call Christine my "harshest critic." But though we don't always agree on matters anthropological, literary, or even philosophical, during our 48 years of marriage I've always known I could rely on Christine for an honest and incisive appraisal of my work and suggestions for improvement. After I retired from academic life in December 2019 (great timing, as it turns out, in view of the pandemic), Christine expected a more relaxed and carefree lifestyle with more freedom to travel. Yet she has patiently tolerated days, including weekends, of me shut in my home study, typing away, so I'm immeasurably grateful for that as well.

Also deserving special mention is my friend John Acorn, a widely published zoologist and award-winning science educator, broadcaster, and colleague at the University of Alberta, who's always been willing to offer expert advice on matters biological. Especially as I've been somewhat critical of biological sciences in places, I should say that John is not only extremely knowledgeable and always responsive and enthusiastic about giving advice. He is also admirably open-minded and wide-ranging in the ideas he is willing to consider. In particular, the first and fourth sections of this book have benefited significantly from his input. At the same time, of course, any perceived shortcomings of the book are my responsibility alone. And this applies to everyone who has commented on its contents.

Thanks are due to a number of anthropological colleagues, both in Canada and abroad, who for nearly two decades now have helped me develop my thinking about "mystery hominoids." Prominent among these have been the late Colin Groves and Deborah Argue, both of the Australian National University. An expert on *Homo floresiensis*, Debbie has carefully reviewed those parts of the book where I explore possible resemblances between Lio descriptions of their ape-men and paleoanthropological interpretations of floresiensis. Again, Debbie and I haven't agreed on all points, but I naturally respect her expertise and know that the book, whatever its merits, has benefited from her kind assistance—not just recently but since we first began corresponding over ten years ago.

On the paleoanthropological front, I'm also grateful to paleoartist Inge Van Noortwijk of Naturalis, the national museum of natural history in Leiden, Holland, who on the basis of our current knowledge of the morphology and behavior of *Homo floresiensis*, has with great care, patience, and enthusiasm composed the splendid illustration of that hominin. The picture of "Floortje" (cf. English "Florrie"), as Inge likes to call her creation, appears in chapter 10, as figure 10.1.

Due to the linear nature of written script, someone has to be thanked last. But as with all mentions, the order in which people appear is no reflection on either their importance or the measure of my gratitude. As I discuss in the book, the topic I address is a challenging one, so I'm especially grateful to my agent, Peter Tallack, who from the beginning has had faith in the project and has proved his support in many ways, not least by carefully editing drafts of the book proposal and, in effect, teaching me to write less like an academic! By the same token, much is owed to my editors at Pegasus—Claiborne Hancock, Jessica Case, and Maria Fernandez—for the proficient, expeditious, and very amicable way in which they've helped me see the book to completion.

Finally, and on a more formal note, I'd like to thank the Social Sciences and Humanities Research Council of Canada for regularly funding my work, on a variety of anthropological topics, over a period of three decades.

—Gregory Forth
Edmonton, Canada

NOTES

CHAPTER 1: Ape-men of Flores Island

1   T. Jacob et al. "Pygmoid Australomelanesian *Homo sapiens* Skeletal Remains from Liang Bua, Flores: Population Affinities and Pathological Abnormalities." *Proceedings of the National Academy of Sciences* 103, no. 36 (2006): 13421-13426.

2   Robin Hemley, *Invented Eden: The Elusive Disputed History of the Tasaday* (Lincoln and London: University of Nebraska Press, 2003), 254, 285n, mentions another species of civet, *Viverricula indica*, which the Tasaday people of the Philippines imagine in a similarly fantastic way. This they describe as looking "more like a fox than a dog," always defecating in the same spot, and jumping on the backs of deer and wild pigs, biting their necks, and eating only their brains.

3   I first referred to the Lio hominoids in Gregory Forth, "Hominids, Hairy Hominoids and the Science of Humanity," *Anthropology Today* 21, no. 3 (June 2005): 13–17. In this and other earlier writings, I transcribed the hominoids' name as "lae ho'a." "Lai ho'a" more accurately reflects the Lio pronunciation.

4   Eriko Aoki, "Piercing the Sky, Cutting the Earth: The Poetics of Knowledge and the Paradox of Power among the Wologai of Central Flores." (PhD diss., The Australian National University, 1996), 227–228.

5   Paul Arndt, *Li'onesisch-Deutsches Wörterbuch* (Ende, Flores: Arnoldus-Druckerei, 1933), 190. (See also pages 54, 60, 80, 136, 171, 432–433, 475, 484, 519).

6   B. C. C. M. M. van Suchtelen, *Endeh* (Weltevreden: Papyrus, 1921), 254, 374.

7   Although previously contested, the Malay origin of "orang utan" as a name for the ape has been demonstrated by Wayan Jarrah Sastrawan, "The Word 'Orangutan': Old Malay Origin or European Concoction?" *Bijdragen tot de Taal-, Land- en Volkenkunde* 176 (2020): 532–541.

8   Mike Morwood and Penny Van Oosterzee, *A New Human: The Startling Discovery and Strange Story of the "Hobbits" of Flores, Indonesia* (New York: Harper-Collins, 2007).

CHAPTER 2: Ape-men as Natural Creatures

1   Walter Auffenburg, *The Komodo Monitor* (Gainesville: University of Florida Press, 1981), 243. G. B. Corbet and J. E. Hill, *The Mammals of the Indomalayan Region* (Oxford: Oxford University Press, 1992), 166, report a head and body length ranging from 41 to 65 centimeters for long-tailed macaques, but these figures come from mainland Southeast Asia and western Indonesia, where specimens are typically larger than on Flores.

2   On the basis of more limited evidence, I previously suggested that Lio ape-man stature might be bimodal, revealing two peaks on a distribution curve. G. Forth, *Images of the Wildman in Southeast Asia: An Anthropological Perspective* (London and New York: Routledge, 2008). The far larger body of information reviewed here does not bear this out.

3   Josef Glinka, "Apakah orang Indonesia bertambah tinggi?" ("Are Indonesians Getting Taller?"), unpublished; H. J. T. Bijlmer, *Outlines of the Anthropology of the Timor Archipelago* (Weltevreden: G. Kolff, 1929); Wilhelmina Keers, *An Anthropological Survey of the Eastern Lesser Sunda Islands* (Amsterdam: Koninklijke Vereeniging Indisch Instituut, 1948).

4   Flores monkey metaphors are discussed in Gregory Forth, *A Dog Pissing at the Edge of a Path* (Montreal and Kingston: McGill-Queens University Press, 2019), 146-159.

5   Forth, *Images*, 266.

6   Forth, *Images*, 156–158.

7   Arndt, *Li'onesisch-Deutsches Worterbuch*, 432–433, s.v. *sala*.

8   On the attribution of tails to African pygmies, see Georg Schweinfurth, *The Heart of Africa*, trans. Ellen E. Frewer (London: Sampson Low, Marston, Low, and Searle, 1873), 2: 137. The rumored tails and dog-like heads of Andaman Islanders are discussed in Clare Anderson, "Writing Indigenous Women's Lives in the Bay of Bengal: Cultures of Empire in the Andaman Islands, 1789-1906," *Journal of Social History* 45, no. 2 (2011): 485; in Kim M. Phillips, *Before Orientalism: Asian Peoples and Cultures in European Travel Writing, 1245-1510* (Philadelphia: University of Pennsylvania Press, 2013), 97; and in Frederic John Mouat, *Adventures and Researches among the Andaman Islanders* (London: Hurst and Blackett Publishers, 1863), 83, 120. Information on Bimanese ideas about the Wawo people of Sumbawa was related to the author, mostly in a series of emails, by their ethnographer, Michael Hitchcock (see further Chapter 8). Polo's original statement attributing dog-like heads to the islanders can be found in *The Book of Ser Marco Polo* (New York: Charles Scribner's Sons, 1903), 2: 309, 311–312.

9   Arndt, *Li'onesisch-Deutsches Wörterbuch*, 27, also gives *binata*, evidently a borrowing from Indonesian "binatang," as a Lio term for "large animal, especially quadruped."

10   Gregory Forth, "Elderly People Growing Tails," *Current Anthropology* 59, no. 4 (2018): 397-414. In his 1933 dictionary Arndt (489, s.v. *tembu*) writes, "According to the Lio, all old people grow a small tail." More recently, Aoki, "Piercing the Sky," 248, reports that the western Lio "talk about old people who had a tail as long as 20 cm." Further afield, the east African Meru people describe a fully human traditional religious functionary (the "mugwe") as possessing a tail. See Bernardo Bernardi, *The Mugwe, a Failing Prophet: A Study of a Religious and Public Dignitary of the Meru of Kenya* (Oxford: Oxford University Press, 1959).

11   "10 Things You Probably Didn't Know About Chimpanzees," issued by the Jane Goodall Institute of Canada, https://janegoodall.ca/our-stories/10-things-about-chimpanzees/. The tenth entry (listed first) is the fact that chimps and other apes do not have tails whereas monkeys do.

12   Differences in fertility rate between cultivators and modern food collectors and between humans, non-sapiens hominins, and apes, owing largely to dietary differences, is discussed in Daniel Lieberman, *The Story of the Human Body: Evolution, Health, and Disease* (New York: Pantheon Books, 2013), 94–95, 96, 187. Given Lio descriptions of ape-men as thin, it may also be worth noting that Lieberman infers non-sapiens hominins, specifically *Homo erectus*, would have been "scrawny."

13   Jinping Yu and F. Stephen Dobson, "Seven Forms of Rarity in Mammals," *Journal of Biogeography* 27 (2000): 131–139. P. G. Harnik, C. Simpson, and J. L. Payne, "Long-term Differences in Extinction Risk among the Seven Forms of Rarity," *Proceedings of the Royal Society* 279 (2012): 4969-4976.

14   Koichi Kitanishi, "Cultivation by the Baka Hunter-Gathers in the Tropical Rain Forest of Central Africa," *African Study Monographs*, Supplement 28 (November 2003): 143-157.

15   G. Forth, *Images*, 21, 75, 192, 193, 198. Roger Bacon, *The Opus Majus of Roger Bacon*, (Philadelphia: University of Pennsylvania Press, 1928. First published in 1268) 2: 387, describes how mystery hominoids in central Asia could be captured by first getting them drunk on beer. R. Bernheimer, *Wild Men in the Middle Ages* (Cambridge, Mass.: Harvard University Press, 1952), 25, mentions the same method of capture as a feature of ancient Italian and German tales about the "wildmen" of Europe.

16   Gregory Forth, *Why the Porcupine is Not a Bird* (Toronto: University of Toronto Press, 2016), 129, 132–133.

17   G. Forth, *Images*, 52, 97, 139. A preference for damp or cloudy weather is mentioned for the *ana ula* hominoids of Rawe (west central Flores), the "mili mongga" of Sumba, and the "orang pendek" of Sumatra. Interestingly, Sumbanese explain the appearance of mili mongga at such times with reference to their feeding on snails, slugs, and other invertebrates, which also emerge in damp conditions.

18   G. Forth, *Images*, 16, 98.

19   D. Falk et al. "The Brain of *Homo floresiensis*," *Science* 308 (2005): 242–245; D. Falk et. al. "Brain Shape in Human Microcephalics and *Homo floresiensis*." *Proceedings of the National Academy of Sciences* 104, no. 7 (2007): 2513–2518. Debbie Argue, "Invited Perspective: The Enigma of *Homo floresiensis*," in *First Islanders*, Peter Bellwood (Hoboken, NJ: Wiley Blackwell, 2017), 60–65.

CHAPTER 3: Ape-men as Supernaturals

1   Arndt, *Li'onesisch-Deutsches Wörterbuch*, 223 s.v. *mae*. On the Lio high god or supreme being lacking a body, also see Paul Arndt, "Dua Nggae, Das Höchste Wesen im Lio-Gebiet (mittel-Flores)," *Annali Laternensi* 3 (1939): 168, 180.

2   Daniel Schmitt, "Compliant Walking in Primates," *Journal of the Zoological Society of London* 284 (1999): 156–157. Hartmut Geyer, et al. "Compliant Leg Behaviour Explains Basic Dynamics of Walking and Running," *Proceedings of the Royal Society B* 273 (2006): 2861–2867. For a demonstration of the Groucho Walk see www.rockysworkouts.com: Groucho Walks. Curiously enough, in *Sasquatch: Legend Meets Science* (New York: Forge Books, 2006), 244, physical anthropologist Jeff Meldrum discusses a reported sighting of the North American mystery hominoid "Bigfoot" where the creature "seemed to glide or float as it moved."

3   Latin names for all plants come from J. A. J. Verheijen, *Dictionary of Plant Names in the Lesser Sunda Islands* (Canberra: Australian National University, 1990).

4   Edem Archibong Eniang, "The Power of the Manatee," *Heroes* 2 (2018): 306; G. T. Basden, *Among the Ibos of Nigeria* (New York: Barnes & Noble, 1966 [1921]), 142–143.

5   Gregory Forth, "Rare Animals as Cryptids and Supernaturals: The Case of Dugongs on Flores Island," *Anthrozoös* 34, no.1 (2021): 61–76.

6   Gregory Forth, "Ethnographic Reports of Freshwater Turtles on Flores Island: The Possibilities of an Undocumented Chelonian Species," *Herpetological Review* 48, no. 2 (2017): 304–310; Gregory Forth, "Further Evidence for Undocumented Freshwater Turtles on Flores Island, Indonesia," *Herpetological Review* 49, no. 1 (2018): 19–21; Gregory Forth, "Ethnographic Evidence for the Presence of the Coconut Crab *Birgus latro* (Linnaeus, 1767) (Anomura, Coenobitidae) on Flores Island, Indonesia," *Crustaceana* 92, no. 8 (2019): 921–941.

7   A description of Javanese beliefs about tuyul (previously written "tujul") is found in Clifford Geertz, *The Religion of Java* (New York: The Free Press, 1960), 16–17.

CHAPTER 4: Ape-men in Myth and Legend

1   G. Forth, *Images*, 12–84.

2   For an example from Vietnam, see G. Forth, *Images*, 169–70. See also Gregory Forth, "Disappearing Wildmen: Capture, Extirpation, and Extinction as Regular Components of Representations of Putative Hairy Hominoids," in *The Anthropology of Extinction: Essays on*

Culture and Species Death, ed. Genese Sodikoff, 200–218 (Bloomington: Indiana University Press, 2012).

3   E. W. Barber and P. T. Barber, *When They Severed Earth from Sky: How the Human Mind Shapes Myth* (Princeton, NJ: Princeton University Press, 2004).

4   Paul Arndt, "Der Kult der Lionesen," *Annali Lateranensi* 8 (1944): 156–157, 161.

CHAPTER 5: What Someone Told Someone Else, or Secondhand Stories of Ape-man Encounters

1   Bijlmer, *Outlines*, 222–227.

2   Instances are described in G. Forth, *Images*, 92, 199, 213.

CHAPTER 6: From Questionable Stories to Compelling Reports

1   A brief summary of the first version of Koli's story (under another pseudonym) appears in G. Forth, *Images*, 67.

2   B. J. Coates and K. D. Bishop, *A Guide to the Birds of Wallacea: Sulawesi, the Moluccas and Lesser Sunda Islands, Indonesia* (Alderley, Australia: Dove Publications, 1997), 248.

3   D. Skrzypińska et al., "Phenomenology of Dream-Reality Confusion: A Quantitative Study," *Dreaming* 28, no. 3 (2018): 245–260.

4   See Sir James Tennet, *Sketches of the Natural History of Ceylon* (London: Longman, Green, Longman and Roberts, 1861), 32, where the relevant civet behavior is described as sucking the blood of domestic fowls leaving "a wound so small as to be almost imperceptible."

5   Pierre Pfeffer, "Notes sur le peuplement mammalian des îles de Florès, Komodo et Rintja (Petites îles de la Sonde)," *Mammalia* 21 (1957): 405–419. Email messages to author from Michael Huffmann, July 5, 2020. Pfeffer's report is based on observations he made of a troop of long-tailed macaques on Rinca Island, just off Flores's western extremity.

CHAPTER 7: More Remarkable Encounters

1   Khitam Al-Refu "Hair Loss in Children, Etiologies and Treatment," *IntechOpen* (2017): 255–267. Rodney P. R. Dawber and Dominique Van Neste, *Hair and Scalp Disorders: Common Presenting Signs, Differential Diagnosis and Treatment*, 2nd ed. (London and New York: Martin Dunitz, 2004), 122–123.

CHAPTER 8: The Little Naked Man and Other Extraterritorial Sightings

1   G. Forth, *Images*, 102–103.

2   G. Forth, *Images*, 102. All information on his encounter with the mystery primates was provided by Hitchcock in a series of emails dating from November–December 2004 to December 2020 and July 2021, during a meeting in London in November 2005, and in copies of Hitchcock's 1982 field notes.

CHAPTER 9: What do Stories of Ape-men Tells Us and What Else Could They Be?

1   M. Pagel and W. Bodmer, "A Naked Ape Would Have Fewer Parasites," *Proceedings of the Royal Society B* 270 (2003): 117–19.

2   Elizabeth Loftus, *Eyewitness Testimony*, 2nd ed. (Cambridge, MA: Harvard University Press, 1979). On the continuing use of eyewitness testimony by courts and the reliability of much of such testimony, see John Kaplan's foreword to this edition of Loftus's book (1979), viii.

3   Goodman and Pas-Alonso, "Trauma and Memory: Normal versus Special Memory Mechanisms," in *Memory and Emotion: Interdisciplinary Perspectives*, ed. Bob Uttl et al. (Oxford:

Blackwell Publishing, 2006), 235; Julia Shaw, *The Memory Illusion: Remembering, Forgetting, and the Science of False Memory* (Toronto: Anchor Canada, 2017).

4　Forth, "Ethnographic Reports of Freshwater Turtles," 304–310; Forth, "Further Evidence for Undocumented Freshwater Turtles," 19–21; Forth, "Ethnographic Evidence for the Presence of the Coconut Crab," 921–941.

5　Shaw, *The Memory Illusion*, 165, 177–180.

6　Goodman and Pas-Alsonso, "Trauma and Memory," 236.

7　Masahiro Mori, "The Uncanny Valley," translated by Karl F. MacDorman and Norri Kageki, *Robotics*, June 12, 2012, https://spectrum.ieee.org/the-uncanny-valley. Laura Spinny, "Exploring the Uncanny Valley: Why Almost-Human Is Creepy," *New Scientist*, October 26, 2016, https://www.newscientist.com/article/mg23230970-500-exploring-the-uncanny-valley-why-almosthuman-is-creepy/#ixzz777OKrVng.

8　P. A. Levine, *Trauma and Memory: Brain and Body in a Search for the Living Past* (Berkeley, CA: North Atlantic Books 2015), 7–9. Levine's main focus is severe, pathological trauma, whereas other authors I cite employ "trauma" in a more inclusive sense, as a reference to experiences causing varying degrees of emotional disturbance.

9　On the "trauma superiority effect," see Shaw, *The Memory Illusion*, 164–166. On the accuracy of emotionally disturbing memory, see also Goodman and Pas-Alonso, "Trauma," 234–235, who refer to research showing that such memories, also called "flashbulb memories," decline slowly over time; also J. F. Kihlstrom, "Trauma and Memory Revisited," in *Memory and Emotion: Interdisciplinary Perspectives*, ed. B. Uttl, et al. (Oxford: Blackwell Publishing, 2006), 259–291, who concludes that "everything we know about memory tells us that emotional involvement makes events more memorable, not less." A riveting critique of therapeutic practices designed to uncover "repressed memories" is found in Elizabeth Loftus and Katherine Ketcham, *The Myth of Repressed Memory: False Memories and Allegations of Sexual Abuse* (New York: St. Martin's Griffin, 1996).

10　Goodman and Pas-Alonso, "Trauma," 234, 237.

11　Email messages to author from Dr. Michael Huffmann, Kyoto University's Primate Research Institute (July 5, 2020), and Professor Linda Fedigan, University of Calgary (June 30, 2020).

12　Corbet and Hill, *The Mammals of the Indomalayan Region*, 166.

13　Bob Holmes, "How Many Uncontacted Tribes are Left in the World?" *New Scientist* (August 31, 2013), https://www.newscientist.com/article/dn24090-how-many-uncontacted-tribes-are-left-in-the-world/.

14　C. Turnbull, *The Forest People: A Study of the Pygmies of the Congo* (New York: Simon and Schuster, 1961), 31; P. McAllister, *Pygmonia: In Search of the Secret Land of the Pygmies* (St. Lucia, Queensland: University of Queensland Press, 2010), 61; J. Glinka, M. D. Artaria, and T. Koesbardiat, "The Three Human Morphotypes in Indonesia," *Indonesian Journal of Social Sciences* 2, no. 2 (2010): 70–76.

15　P. Schebesta, *Among Congo Pigmies*, trans. Gerald Griffin (London: Hutchinson & Co., 1933), 40; *Die Negrito Asiens* (Wien-Mödling: St. Gabriel-Verlag,1957), 2: 58; *Among the Forest Dwarfs of Malaya* (London: Hutchinson & Co. Ltd., 1929), 216–17. McAllister, *Pygmonia*, 11, 68, suggests that the Malay name for neighboring negritos, "semang," has been influenced by "siamang," the Malay name, now adopted in English, for the large "local black gibbons." The two words, however, appear to derive from different roots. See Teckwyn Lim, "An Aslian Origin for the Word *Gibbon*," *Lexis* 15 (2020): 10–11.

16　Schweinfurth, *The Heart of Africa*, 137; Edward Tyson, *Orang-outang, sive, Homo sylvestris, or, The Anatomy of a Pygmie Compared with that of a Monkey, an Ape, and a Man* (London: Thomas Bennet, 1699). On the confusion of orangutans and other apes, see also Robert Cribb, Helen Gilbert, and Helen Tiffin, *Wild man from Borneo: A Cultural History of the Orangutan*. (Honolulu: University of Hawai'i Press, 2014).

17   On pygmies stealing from taller cultivators, see Kitanishi, "Cultivation," 143–157. On thieving chimpanzees, see Sabrina Krief et al., "Wild Chimpanzees on the Edge: Nocturnal Activities in Croplands," *PLoS One* 9, no. 10 (2014): 1–11.

18   K. Endicott, *An Analysis of Malay Magic* (Kuala Lumpur: Oxford University Press, 1970), 80–81. On Philippine negritos as spiritual beings, see J. L. Sodusta, *Jamoyawon Ritual: A Territorial Concept* (Quezon City: University of the Philippines, 1983), 51–53, 65, 76–78. On African pygmies: Schweinfurth, *The Heart of Africa*, 136, 141–44; T. Schadeberg, "Batwa: The Bantu Name for the Invisible People," in *Central African Hunter-gatherers in a Multidisciplinary Perspective: Challenging Elusiveness*, eds. Karen Biesbrouck, Stefan Elders, and Gerda Rossel (Leiden: Research School CNWS, 1999), 21–39. In other places and at other times, a similar ability to "vanish" has been attributed by Europeans to North American natives and by the British to South African Boers. See Aidan Forth, *Barbed-wire Imperialism: Britain's Empire of Camps, 1876-1903* (Oakland, CA: University of California Press, 2017), 134.

19   Hemley, *Invented Eden*.

20   Gregory Forth, "Verhoeven's Living Negritos and the Story of Zakharias Ze: A Prehistory of *Homo floresiensis*," *Bijdragen tot de Taal-, Land- en Volkenkunde* (2022) forthcoming. Jacob et al. "Pygmoid Australomelanesian *Homo sapiens* Skeletal Remains." S. Tucci et al. "Evolutionary History and Adaptation of a Human Pygmy Population on Flores Island, Indonesia," *Science* 361 (2018): 511–16.

21   J. Glinka, "Apakah orang Indonesian bertambah tinggi?" Bijlmer, *Outlines*.

CHAPTER 10: Could Ape-men Survive on Flores Island?

1   Morwood and Van Oosterzee, *A New Human*, 154–155. Gregory Forth, *Beneath the Volcano: Religion, Cosmology, and Spirit Classification among the Nage of Eastern Indonesia* (Leiden: KITLV Press, 1998), 154–155.

2   Morwood and Van Oosterzee, *A New Human*, 105. On the wrists see: M. W. Tocheri et al., "The Primitive Wrist of *Homo floresiensis* and its Implications for Hominin Evolution," *Science* 317 (September 21, 2007): 1743–1745.

3   S.G. Larson et al., "*Homo floresiensis* and the Evolution of the Hominin Shoulder," *Journal of Human Evolution* 53 (2007): 718–731.

4   Morwood and Van Oosterzee, *A New Human*, 105.

5   Larson et al., "*Homo floresiensis*," 729.

6   W. L. Jungers et al., "The Foot of *Homo floresiensis*," *Nature* 459 (2009): 81–84; W. L. Jungers et al., "Descriptions of the Lower Limb Skeleton of *Homo floresiensis*," *Journal of Human Ecology* 57 (2009): 538–554.

7   M. Zannis, "So, You Walk Like a Duck: The Evolution of the Collapsed Arch," July 19, 2016. https://www.medbridgeeducation.com/blog/2016/07/. Sachini N. K. Kodithuwakku Arachchige et al., "Flatfeet: Biomechanical Implications, Assessment and Management," *The Foot* 38 (2019): 81–85.

8   Ewen Calloway, "Flores 'Hobbit' Walked more like a Clown than Frodo," *New Scientist* 53 (April 16, 2008): 983–984.

9   H. Van Werkhoven and S. J. Piazza, "Foot Structure is Correlated with Performance in a Single-joint Jumping Task," *Journal of Biomechanics* 57 (2017): 27–31.

10   Bellwood, *First Islanders*, 65–67. On evidence for fire: M. W. Morley et al., "Initial Micromorphological Results from Liang Bua, Flores (Indonesia): Site Formation Processes and Hominin Activities at the Type Locality of *Homo floresiensis*," *Journal of Archaeological Science* 77 (2016): 125–42.

11   Morwood and Van Oosterzee, *A New Human*, 177; Falk et al. "The Brain of *Homo floresiensis*," 242–245. Falk et al. "Brain Shape in Human Microcephalics," 2513–2518.

12 P. Brown and T. Maeda. "Liang Bua *Homo floresiensis* Mandibles and Mandibular Teeth: A
    Contribution to the Comparative Morphology of a New Hominin Species," *Journal of Human
    Evolution* 57 (2009): 571–596.

13 P. T. Brown, T. Sutikna, et al. "A New Small-bodied Hominin from the Late Pleistocene of
    Flores, Indonesia," *Nature* 431 (2004): 1055–1061. Debbie Argue, email to author, August 1,
    2021.

14 T. Sutikna et al., "Revised Stratigraphy and Chronology for *Homo floresiensis* at Liang Bua in
    Indonesia," *Nature* 532 (2016): 366–369.

15 G. van den Bergh et al., "*Homo floresiensis*-like Fossils from the Early Middle Pleistocene of
    Flores," *Nature* 534 (2016): 245–248. A. Brumm et al., "Hominins on Flores, Indonesia, by One
    Million Years Ago," *Nature* 464 (2010): 748–752.

16 A. van der Geer et al., *Evolution of Island Mammals: Adaptation and Extinction of Placental
    Mammals on Islands* (Chichester, England: Wiley-Blackwell, 2012), 194. Forth, *Why the
    Porcupine*, 322–323.

17 Morwood and van Oosterzee, *A New Human*, 154–155. Sutikna et al. "Revised Stratigraphy," 366.

18 Richard Dawkins, "Just Don't Call Them Hobbits," *Los Angeles Times*, November 9, 2004.
    https://www.latimes.com/archives/la-xpm-2004-nov-09-oe-dawkins9-story.html.

19 J. Maringer and Th. Verhoeven, "Die Steinartefakte aus der Stegodon-Fossilschicht von
    Mengeruda auf Flores, Indonesien," *Anthropos* 65 (1970): 229–247.

20 Ewen Calloway, "Hobbit Relatives Hint at Family Tree," *Nature* 534 (2016): 164–165.

21 Bill Bryson, *A Short History of Nearly Everything* (New York: Doubleday, 2003): 390–391.
    See also Richard Leakey and Roger Lewin, *The Sixth Extinction: Biodiversity and Its Survival*
    (London: Weidenfeld & Nicholson, 1996).

22 G. D. Van den Bergh et al., "The Liang Bua Faunal Remains: A 9.5 k.yr. Sequence from Flores,
    East Indonesia," *Journal of Human Evolution* 57 (2009): 527–537.

23 R. Dennell, et al., "The Origins and Persistence of *Homo floresiensis* on Flores: Biogeographical
    and Ecological Perspectives," *Quaternary Science Reviews* 96 (2014): 98–107.

24 Jason Daley, "Ancient Humans Weathered the Toba Supervolcano Just Fine," *Smithsonian
    Magazine*, March 14, 2018, https://www.smithsonianmag.com/smart-news/ancient-humans
    -weathered-toba-supervolcano-just-fine-180968479/.

25 M. Morwood et al. "Archaeology and Age of a New Hominin from Flores in Eastern
    Indonesia," *Nature* 431 (2004): 1087–1091; Morwood and van Oosterzee, *A New Human*, 238.

26 "Red Deer Cave Thigh Bone Suggests Prehistoric Humans Survived Until Recently,"
    Australian Broadcasting Corporation, December 21, 2015, https://www.cbc.ca/news/science
    /red-deer-cave-people-1.3374837. Curnoe D. et al., "A Hominin Femur with Archaic Affinities
    from the Late Pleistocene of Southwest China," *PLoS ONE* 10, no. 12 (December 17, 2015):
    e01433320143332.

27 Information on dates for food production comes from Bellwood, *First Islanders*, and Peter
    Bellwood, email to author, May 10, 2021. Dates for pigs, dogs, and other animals introduced by
    humans are from Van den Bergh et al., "The Liang Bua Fossil Remains," 527–537.

28 Van den Bergh et al., "*Homo floresiensis*-like Fossils," 245.

29 "Curiouser and curiouser: Coelacanths are even weirder than previously thought," *The
    Economist*, June 17, 2021, citing K. Mahé et al. "New scale analyses reveal centenarian African
    coelacanths," *Current Biology* 31(16): 3621–3628.

30 David Biello, "Laotian Rodent Proves Living Fossil," *Scientific American*, March 10, 2006,
    https://www.scientificamerican.com/article/laotian-rodent-proves-liv/. Susan L. Jewett,
    "On the Trail of the Coelacanth, a Living Fossil," *The Washington Post*, November 11, 1998,
    https://www.washingtonpost.com/wp-srv/national/horizon/nov98/fishstory.htm.

31 Bellwood, *First Islanders*, 58.

32 Sutikna et al., "Revised Stratigraphy," 366.

33 Pieter J. den Boer, "The Present Status of the Competitive Exclusion Principle," *Trends in Ecology & Evolution* 1 (1986): 25–28.

34 Morwood and Van Oosterzee, *A New Human*, 146.

35 Georges Cuvier, *Recherches sur les ossemens fossils de quadrupeds* (Cambridge: Cambridge University Press, 2015 [1812]).

36 Scott Weidensaul, *The Ghost with Trembling Wings: Science, Wishful Thinking, and the Search for Lost Species* (New York: North Point Press, 2002).

37 Exemplifying a cognitive approach is Pascal Boyer's work on religion, e.g., *The Naturalness of Religious Ideas: A Cognitive Theory of Religion* (Berkeley: University of California Press, 1994). The older thesis concerning anomalous animals as supernaturally powerful is most closely associated with Mary Douglas, *Purity and Danger: An Analysis of Concepts of Pollution and Taboo* (London: Routledge, 2002 [1966]).

38 As good an example as any is G. G. Simpson, "Mammals and Cryptozoology," *Proceedings of the American Philosophical Society* 128 (1984): 1–19.

39 Although ontological pluralists come in several varieties, in regard to how different people might conceive of similarities and differences between humans and animals, among the best known is Philippe Descola, *Beyond Nature and Culture*, trans. Janet Lloyd (Chicago: University of Chicago Press, 2013).

40 Originally suggested by the present author to Martin Walsh, this concept of "shadow" has been applied by Walsh and Helle Goldman, "Cryptids and Credulity: The Zanzibar Leopard and other Imaginary Beings," in *Anthropology and Cryptozoology*, ed. Samantha Hurn (London and New York: Routledge, 2017), 84. Their chapter concerns recent reports of a subspecies of leopard (*Panthera pardus adersi*), officially extinct but still occasionally reported on Zanzibar.

41 Michael Ruse, *The Philosophy of Human Evolution* (Cambridge: Cambridge University Press, 2012), 102–107.

42 Henry Gee, "Our Not So Distant Relative," *The Guardian*, 28 (October 2004). Morwood and Van Oostersee, *A New Human*, 146, give a slightly different version of Brown's statement. With reference to the discovery of *Homo floresiensis* and especially the geologically recent dates, they cite Brown as saying "if a spacecraft had landed in a rice paddy and deposited an alien, it would have been easier for him to accept."

BIBLIOGRAPHY

Al-Refu, Khitam. "Hair Loss in Children, Etiologies and Treatment." *IntechOpen* (2017).

Anderson, Clare. "Writing Indigenous Women's Lives in the Bay of Bengal: Cultures of Empire in the Andaman Islands, 1789–1906." *Journal of Social History* 45, no. 2 (2011): 480–496.

Aoki, Eriko. "Piercing the Sky, Cutting the Earth: The Poetics of Knowledge and the Paradox of Power among the Wologai of Central Flores." PhD diss., The Australian National University, 1996.

Arachchige, Sachini N. K. Kodithuwakku, Harish Chander, and Adam Knight. "Flatfeet: Biomechanical Implications, Assessment and Management." *The Foot* 38 (2019): 81–85.

Argue, Debbie. "Invited Perspective: The Enigma of *Homo floresiensis*." In *First Islanders: Prehistory and Human Migration in Island Southeast Asia*, Peter Bellwood, 60–65. Hoboken, NJ: Wiley Blackwell, 2017.

Arndt, Paul. *Li'onesisch-Deutsches Wörterbuch*. Ende: Arnoldus-Druckerei, 1933.

Arndt, Paul. "Dua Nggae, Das Höchste Wesen im Lio-Gebiet (mittel-Flores)." *Annali Laternensi* 3 (1939): 142–210,

Arndt, Paul. "Der Kult der Lionesen." *Annali Lateranensi* 8 (1944): 155–182.

Auffenburg, Walter. *The Komodo Monitor*. Gainesville: University of Florida Press, 1981.

Bacon, Roger. *The Opus Majus of Roger Bacon*. Volume 2. Philadelphia: University of Pennsylvania Press, 1928 (First published 1268).

Barber, E. W. and P. T. Barber. *When They Severed Earth from Sky: How the Human Mind Shapes Myth*. Princeton, NJ: Princeton University Press, 2004.

Basden, G. T. *Among the Ibos of Nigeria.* New York: Barnes & Noble, 1966 (First published 1921).

Bellwood, Peter. *First Islanders: Prehistory and Human Migration in Island Southeast Asia.* Hoboken, NJ: Wiley Blackwell, 2017.

Bernardi, Bernardo. *The Mugwe, A Failing Prophet: A Study of a Religious and Public Dignitary of the Meru of Kenya.* Oxford: Oxford University Press, 1959.

Bernheimer, Richard. *Wild Men in the Middle Ages: A Study in Art, Sentiment, and Demonology.* Cambridge, MA.: Harvard University Press, 1952.

Biello, David. "Laotian Rodent Proves Living Fossil." *Scientific American.* March 10, 2006. https://www.scientificamerican.com/article/laotian-rodent-proves-liv/.

Bijlmer, H. J. T. *Outlines of the Anthropology of the Timor Archipelago.* Weltevreden: G. Kolff, 1929.

Boyer, Pascal. *The Naturalness of Religious Ideas: A Cognitive Theory of Religion.* Berkeley: University of California Press, 1994.

Brown, P., T. Sutikna, M. J. Morwood, R. P. Soejono, Jatmiko, E. Wayhu Saptomo, and Rokus Due Awe. "A New Small-bodied Hominin from the Late Pleistocene of Flores, Indonesia." *Nature* 431 (2004): 1055–1061.

Brown, P. T. and T. Maeda. "Liang Bua *Homo floresiensis* Mandibles and Mandibular Teeth: A Contribution to the Comparative Morphology of a New Hominin Species." *Journal of Human Evolution* 57 (2009): 571–596.

Brumm, Adam, Gitte M. Jensen, Gert D. van den Bergh, Michael J. Morwood, Iwan Kurniawan, Fachroel Aziz, and Michael Storey. "Hominins on Flores, Indonesia, by One Million Years Ago." *Nature* 464 (2010): 748–752.

Bryson, Bill. *A Short History of Nearly Everything.* New York: Doubleday, 2003.

Calloway, Ewen. "Flores 'hobbit' walked more like a clown than Frodo." *New Scientist* 53 (April 16, 2008): 983–984.

Calloway, Ewen. "Hobbit Relatives Hint at Family Tree." *Nature* 534 (2016): 164–165.

Coates, B. J. and K. D. Bishop. *A Guide to the Birds of Wallacea: Sulawesi, the Moluccas and Lesser Sunda Islands.* Alderley, Australia: Dove Publications, 1997.

Corbet, G. B. and J. E. Hill. *The Mammals of the Indomalayan Region: A Systematic Review.* Oxford: Oxford University Press, 1992.

Cribb, Robert, Helen Gilbert, and Helen Tiffin. *Wild Man from Borneo: A Cultural History of the Orangutan.* Honolulu: University of Hawai'i Press, 2014.

Curnoe, Darren, Xueping Ji, Wu Liu, Zhende Bao, Paul S. C. Taçon, and Liang Ren. "A Hominin Femur with Archaic Affinities from the Late Pleistocene of Southwest China." *PLoS ONE*, 10, no. 12 (2015): e01433320143332.

Cuvier, Georges. *Recherches sur les ossemens fossils de quadrupeds*. Cambridge: Cambridge University Press, 2015 (First published 1812).

Daley, Jason. "Ancient Humans Weathered the Toba Supervolcano Just Fine." *Smithsonian Magazine*. March 14, 2018. https://www.smithsonianmag .com/smart-news/ancient-humans-weathered-toba-supervolcano-just-fine -180968479/.

Dawber, Rodney P. R. and Dominique Van Neste. *Hair and Scalp Disorders: Common Presenting Signs, Differential Diagnosis and Treatment*. Second Edition. London and New York: Martin Dunitz, 2004.

Dawkins, Richard. "Just Don't Call Them Hobbits." *Los Angeles Times*, November 9, 2004. https://www.latimes.com/archives/la-xpm-2004-nov-09-oe-dawkins9 -story.html.

Den Boer, P. J. "The Present Status of the Competitive Exclusion Principle." *Trends in Ecology & Evolution* 1 (1986): 25–28.

Dennell, Robin W., Julien Louys, Hannah J. O'Regan, and David M. Wilkinson. "The Origins and Persistence of *Homo floresiensis* on Flores: Biogeographical and Ecological Perspectives." *Quaternary Science Reviews* 96 (2014): 98–107.

Descola, Philippe. *Beyond Nature and Culture*. Translated by Janet Lloyd. Chicago: University of Chicago Press, 2013.

Douglas, Mary. *Purity and Danger: An Analysis of Concepts of Pollution and Taboo*. London: Routledge, 2002 (First published 1966).

Endicott, Kirk. *An Analysis of Malay Magic*, Kuala Lumpur: Oxford University Press, 1970.

Eniang, Edem Archibong. "The Power of the Manatee." *Heroes* 2 (2018): 306.

Falk, Dean, Charles Hildebolt, Kirk Smith, M. J. Morwood, Thomas Sutikna, Peter Brown, Jatmiko, E. Wayhu Saptomo, Barry Brunsden, and Fred Prior. "The Brain of *Homo floresiensis*." *Science* 308 (2005): 242–245.

Falk, Dean, Charles Hildebolt, Kirk Smith, M. J. Morwood, Thomas Sutikna, Jatmiko, E. Wayhu Saptomo, Herwig Imhof, Horst Seidler, and Fred Prior. "Brain Shape in Human Microcephalics and *Homo floresiensis*." *Proceedings of the National Academy of Sciences* 104, no. 7 (2007): 2513–2518.

Forth, Aidan. *Barbed-wire Imperialism: Britain's Empire of Camps, 1876-1903*. Oakland: University of California Press, 2017.

Forth, Gregory. *Beneath the Volcano: Religion, Cosmology, and Spirit Classification Among the Nage of Eastern Indonesia*. Leiden: KITLV Press, 1998.

Forth, Gregory. "Hominids, Hairy Hominoids and the Science of Humanity." *Anthropology Today* 21, no. 3 (2005): 13–17.

Forth, Gregory. *Images of the Wildman in Southeast Asia: An Anthropological Perspective*. London and New York: Routledge, 2008.

Forth, Gregory. "Disappearing Wildmen: Capture, Extirpation, and Extinction as Regular Components of Representations of Putative Hairy Hominoids." In *The Anthropology of Extinction: Essays on Culture and Species Death*, edited by Genese Sodikoff, 200–218. Bloomington: Indiana University Press, 2012.

Forth, Gregory. *Why the Porcupine is Not a Bird: Explorations in the Folk Zoology of an Eastern Indonesian People*. Toronto: University of Toronto Press, 2016.

Forth, Gregory. "Ethnographic Reports of Freshwater Turtles on Flores Island: The Possibilities of an Undocumented Chelonian Species." *Herpetological Review* 48, no. 2 (2017): 304–310.

Forth, Gregory. "Elderly People Growing Tails." *Current Anthropology* 59, no. 4 (2018): 397–414.

Forth, Gregory. "Further Evidence for Undocumented Freshwater Turtles on Flores Island, Indonesia." *Herpetological Review* 48, no. 1 (2018): 19–21.

Forth, Gregory. "Ethnographic Evidence for the Presence of the Coconut Crab *Birgus latro* (Linnaeus, 1767) (Anomura, Coenobitidae) on Flores Island, Indonesia." *Crustaceana* 92, no. 8 (2019): 921–941.

Forth, Gregory. *A Dog Pissing at the Edge of a Path: Animal Metaphors in an Eastern Indonesian Society*. Montreal and Kingston: McGill-Queens University Press, 2019.

Forth, Gregory. "Rare Animals as Cryptids and Supernaturals: The Case of Dugongs on Flores Island." *Anthrozoös* 34, no. 1 (2021): 61–76.

Forth, Gregory. "Verhoeven's Living Negritos and the Story of Zakharias Ze: A Prehistory of *Homo floresiensis*," *Bijdragen tot de Taal-, Land- en Volkenkunde* (2022) forthcoming.

Gee, Henry. "Our Not So Distant Relative." *The Guardian*, October 28, 2004.

Geertz, Clifford. *The Religion of Java*. New York: The Free Press, 1960.

Geyer, Hartmut, Andre Seyfarth, and Reinhard Blickhan. "Compliant Leg Behaviour Explains Basic Dynamics of Walking and Running." *Proceedings of the Royal Society B* 273 (2006): 2861–2867.

Glinka, Josef. "Apakah orang Indonesia bertambah tinggi?" ("Are Indonesians Getting Taller?"). Unpublished.

Glinka, Josef, M. D. Artaria, and T. Koesbardiat. "The Three Human Morphotypes in Indonesia." *Indonesian Journal of Social Sciences* 2, no. 2 (2010): 70–76.

Goodman, G.S. and P. M. Pas-Alonso. "Trauma and Memory: Normal versus Special Memory Mechanisms." In *Memory and Emotion: Interdisciplinary Perspectives*, edited by Uttl, Bob, Nobuo Ohta, and Amy L. Siegenthaler, 233–258. Oxford: Blackwell Publishing, 2006.

Harnik, P. G., C. Simpson, and J. L. Payne. "Long-term Differences in Extinction Risk among the Seven Forms of Rarity." *Proceedings of the Royal Society* 279 (2012): 4969–4976.

Hemley, Robin. *Invented Eden: The Elusive Disputed History of the Tasaday.* Lincoln and London: University of Nebraska Press, 2003.

Holmes, Bob. "How Many Uncontacted Tribes are Left in the World?" *New Scientist.* August 22, 2013. https://www.newscientist.com/article /dn24090-how-many-uncontacted-tribes-are-left-in-the-world/.

Jacob, T., E. Indriati, R. P. Soejono, K. Hsü, D. W. Frayer, R. B. Eckhardt, A. J. Kuperavage, A. Thorne, and M. Henneberg. "Pygmoid Australomelanesian *Homo sapiens* Skeletal Remains from Liang Bua, Flores: Population Affinities and Pathological Abnormalities." *Proceedings of the National Academy of Sciences* 103, no.36 (2006): 13421–13426.

Jewett, Susan L. "On the Trail of the Coelacanth, a Living Fossil." *The Washington Post*, November 11, 1998. https://www.washingtonpost.com/wp-srv/national /horizon/nov98/fishstory.htm.

Jungers, W. L., W. E. H. Harcourt-Smith, R. E. Wunderlich, M. W. Tocheri, S. G. Larson, T. Sutikna, Rhokus Due Awe, and M. J. Morwood. "The Foot of *Homo floresiensis.*" *Nature* 459 (2009): 81–84.

Jungers, W. L., S. G. Larson, W. Harcourt-Smith, M. J. Morwood, T. Sutikna, Rokhus Due Awe, and T. Djubiantonoe. "Descriptions of the Lower Limb Skeleton of *Homo floresiensis.*" *Journal of Human Ecology* 57 (2009): 538–554.

Keers, Wilhelmina. *An Anthropological Survey of the Eastern Lesser Sunda Islands.* Amsterdam: Koninklijke Vereeniging Indisch Instituut, 1948.

Kihlstrom, J. F. "Trauma and Memory Revisited." In *Memory and Emotion: Interdisciplinary Perspectives*, edited by Uttl, Bob, Nobuo Ohta, and Amy L. Siegenthaler, 259–291. Oxford: Blackwell Publishing, 2006.

Kitanishi, Koichi. "Cultivation by the Baka Hunter-Gathers in the Tropical Rain Forest of Central Africa." *African Study Monographs*, Supplement 28 (2003): 143–157.

Krief, Sabrina, Marie Cibot, Sarah Bortolamiol, Andrew Seguya, Jean-Michel Krief, Shelly Masi. "Wild Chimpanzees on the Edge: Nocturnal Activities in Croplands." *PLoS One*, 9, no. 10 (2014): 1–11.

Larson, S. G., William L. Jungers, Michael J. Morwood, Thomas Sutikna, Jatmiko, E. Wahyu Saptomo, Rokus Due Awe, and Tony Djubiantono. "*Homo floresiensis* and the Evolution of the Hominin Shoulder." *Journal of Human Evolution* 53 (2007): 718–731.

Leakey, R. and Roger Lewin. *The Sixth Extinction: Biodiversity and its Survival.* London: Weidenfeld & Nicholson, 1996.

Levine, P. A. *Trauma and Memory: Brain and Body in a Search for the Living Past.* Berkeley, CA: North Atlantic Books, 2015.

Lieberman, Daniel E. *The Story of the Human Body: Evolution, Health, and Disease.* New York: Pantheon Books, 2013.

Lim, Teckwyn. "An Aslian Origin for the Word *Gibbon*." *Lexis* 15 (2020): 1–21.

Loftus, Elizabeth. *Eyewitness Testimony.* Second Edition. Cambridge, MA: Harvard University Press, 1979.

Loftus, Elizabeth and Katherine Ketcham. *The Myth of Repressed Memory: False Memories and Allegations of Sexual Abuse.* New York: St. Martin's Griffin, 1996.

Mahé, Kélig, Bruno Emande, and Marc Herbin. "New scale analyses reveal centenarian African coelacanths," *Current Biology* 31(16): 3621–3628.

Maringer, J. and Th. Verhoeven. "Die Steinartefakte aus der Stegodon-Fossilschicht von Mengeruda auf Flores, Indonesien." *Anthropos* 65 (1970): 229–247.

McAllister, Peter. *Pygmonia: In Search of the Secret Land of the Pygmies.* St. Lucia, Queensland: University of Queensland Press, 2010.

Meldrum, Jeff. *Sasquatch: Legend Meets Science.* New York: Forge Books, 2006.

Mori, Masahiro. "The Uncanny Valley." Translated by Karl F. MacDorman and Norri Kageki. *Robotics* (June 12, 2012). https://spectrum.ieee.org/the-uncanny -valley.

Morley, Mike W., Paul Goldberg, Thomas Sutikna, Matthew W. Tocheri, Linda C. Prinsloo, Jatmiko, E. Wahyu Saptomo, Sri Wasisto, and Richard G. Roberts. "Initial Micromorphological Results from Liang Bua, Flores (Indonesia): Site Formation Processes and Hominin Activities at the Type Locality of *Homo floresiensis.*" *Journal of Archaeological Science* 77 (2016): 125–142.

Morwood, M. J., R. P. Soejono, R. G. Roberts, T. Sutikna, C. S. M. Turney, K. E. Westaway, W. J. Rink, et al. "Archaeology and Age of a New Hominin from Flores in Eastern Indonesia." *Nature* 431 (2004): 1087–1091.

Morwood, Mike, and Penny Van Oosterzee. *A New Human: The Startling Discovery and Strange Story of the "Hobbits" of Flores, Indonesia.* New York: Harper-Collins, 2007.

Mouat, Frederic John. *Adventures and Researches among the Andaman Islanders.* London: Hurst and Blackett Publishers, 1863.

Pagel, M. and W. Bodmer. "A Naked Ape Would Have Fewer Parasites." *Proceedings of the Royal Society B* 270 (2003): 117–119.

Pfeffer, P. "Notes sur le peuplement mammalian des îles de Florès, Komodo et Rintja (Petites îles de la Sonde)." *Mammalia* 21 (1957): 405–419.

Phillips, Kim M. *Before Orientalism: Asian Peoples and Cultures in European Travel Writing, 1245–1510*. Philadelphia: University of Pennsylvania Press, 2013

Polo, Marco. *The Book of Ser Marco Polo*. Volume 2. Edited by Henry Yule. New York: Charles Scribner's Sons, 1903.

Ruse, Michael. *The Philosophy of Human Evolution*. Cambridge: Cambridge University Press, 2012.

Sastrawan, Wayan Jarrah. "The Word 'Orangutan': Old Malay Origin or European Concoction?" *Bijdragen tot de Taal-, Land- en Volkenkunde* 176 (2020): 532–541.

Schadeberg, T. "Batwa: The Bantu Name for the Invisible People." In *Central African Hunter-gatherers in a Multidisciplinary Perspective: Challenging Elusiveness*, edited by Karen Biesbrouck, Stefan Elders, and Gerda Rossel, 21–39. Leiden: Research School CNWS, 1999.

Schebesta, Paul. *Among the Forest Dwarfs of Malaya*. London: Hutchinson & Co. Ltd., 1929.3Schmitt, Daniel. "Compliant Walking in Primates." *Journal of the Zoological Society of London*. 284 (1999): 149–160.

Schweinfurth, Georg. *The Heart of Africa*. Volume 2. Translated by Ellen E. Frewer. London: Sampson Low, Marston, Low, and Searle, 1873.

Shaw, Julia. *The Memory Illusion: Remembering, Forgetting, and the Science of False Memory*. Toronto: Anchor Canada, 2017.

Simpson, G. G. "Mammals and Cryptozoology." *Proceedings of the American Philosophical Society* 128 (1984): 1–19.

Skrzypińska, Dagna, Małgorzata Hołda, Barbara Szmigielska, and Monika Słodka. "Phenomenology of Dream-Reality Confusion: A Quantitative Study." *Dreaming* 28, no. 3 (2018): 245–260.

Sodusta, Jesucita L. *Jamoyawon Ritual: A Territorial Concept*. Quezon City: University of the Philippines Press, 1983.

Spinney, Laura. "Exploring the Uncanny Valley: Why Almost-human is Creepy." *New Scientist* 3097 (October 26, 2016). https://www.newscientist.com/article /mg23230970-500-exploring-the-uncanny-valley-why-almosthuman-is -creepy/#ixzz777OKrVng.

Sutikna, Thomas, Matthew W. Tocheri, Michael J. Morwood, E. Wahyu Saptomo, Jatmiko, Rokus Due Awe, Sri Wasisto, et al. "Revised Stratigraphy and Chronology for *Homo floresiensis* at Liang Bua in Indonesia." *Nature* 532 (2016): 366–369.

Tennet, Sir James. *Sketches of the Natural History of Ceylon*. London: Longman, Green, Longman and Roberts, 1861.

Tocheri, Matthew W., Caley M. Orr, Susan G. Larson, Thomas Sutikna, Jatmiko, E. Wahyu Saptomo, Rokus Due Awe, Tony Djubiantono, Michael J. Morwood, and William L. Jungers. "The Primitive Wrist of *Homo floresiensis* and its Implications for Hominin Evolution." *Science* 317 (2007): 1743–1745.

Tucci, Serena, Samuel H. Vohr, Rajiv C. McCoy, Benjamin Vernot, Matthew R. Robinson, Chiara Barbieri, Brad J. Nelson, et al. "Evolutionary History and Adaptation of a Human Pygmy Population on Flores Island, Indonesia." *Science* 361 (2018): 511–516.

Turnbull, Colin. *The Forest People: A Study of the Pygmies of the Congo.* New York: Simon and Schuster, 1961.

Tyson, Edward. *Orang-Outang, Sive, Homo Sylvestris, or, The Anatomy of a Pygmie Compared with That of a Monkey, an Ape, and a Man: Wherein It Will Appear That They Are All Either Apes or Monkeys, and Not Men, as Formerly Pretended.* London: Thomas Bennet, 1699.

Van den Bergh, G. D., H. J. M. Meijer, Rokhus Due Awe, M. J. Morwood, K. Szabó, L. W. van den Hoek Ostende, T. Sutikna, E. W. Saptomo, P. J. Piper, and K. M. Dobney. "The Liang Bua Faunal Remains: A 9.5. k.yr. Sequence from Flores, East Indonesia." *Journal of Human Evolution* 57 (2009): 527–537.

Van den Bergh, G. D., Y. Yousuke Kaifu, Iwan Kurniawan, Reiko T. Kono, Adam Brumm, Erick Setiyabudi, Fachroel Aziz, and Michael J. Morwood. "*Homo floresiensis*-like Fossils from the Early Middle Pleistocene of Flores." *Nature* 534 (2016): 245–248.

Van der Geer, A. George Lyras, John de Vos, and Michael Dermitzakis. *Evolution of Island Mammals: Adaptation and Extinction of Placental Mammals on Islands.* Chichester: Wiley-Blackwell, 2012.

Van Suchtelen, B. C. C. M. M. *Endeh (Flores). Mededeelingen van het Bureau voor de Bestuurszaken der Buitengewesten, bewerkt door het Encylopaedisch Bureau.* Aflevering 26. Weltevreden: Papyrus, 1921.

Van Werkhoven, H. and S. J. Piazza. "Foot Structure is Correlated with Performance in a Single-joint Jumping Task." *Journal of Biomechanics* 57 (2017): 27–31.

Verheijen, J. A. J. *Dictionary of Plant Names in the Lesser Sunda Islands.* Pacific Linguistics Series D-No. 83. Canberra: Australian National University, 1990.

Walsh, Martin and Helle Goldman. "Cryptids and Credulity: The Zanzibar Leopard and other Imaginary Beings," in *Anthropology and Cryptozoology: Exploring Encounters with Mysterious Creatures,* edited by Samantha Hurn, 54–90. London and New York: Routledge, 2017.

Weidensaul, Scott. *The Ghost with Trembling Wings: Science, Wishful Thinking, and the Search for Lost Species.* New York: North Point Press, 2002.

Yu, Jinping and F. Stephen Dobson. "Seven Forms of Rarity in Mammals." *Journal of Biogeography* 27 (2000): 131-139.

Zannis, Matthew. "So, You Walk Like a Duck: The Evolution of the Collapsed Arch." July 19, 2016 (Accessed September 3, 2021). https://www.med bridgeeducation.com/blog/2016/07/.

INDEX